Agriculture and Industrialization

THE NATURE OF INDUSTRIALIZATION

Series editors: *Peter Mathias and John A. Davis*

This series is based on graduate seminars in economic history that were held at the University of Warwick's Centre for Social History and sponsored by the *Istituto Italiano per gli Studi Filosofici* (Naples)

Published

Volume 1 *The First Industrial Revolutions*
Volume 2 *Innovation and Technology in Europe*
Volume 3 *Enterprise and Labour*
Volume 4 *Agriculture and Industrialization*

In preparation

Volume 5 *International Trade and British Economic Growth*

The Nature of Industrialization

Edited by
Peter Mathias and John A. Davis

Volume 4

Agriculture and Industrialization:
from the Eighteenth Century to the Present Day

First published 1996

2 4 6 8 10 9 7 5 3 1

Blackwell Publishers Ltd
108 Cowley Road
Oxford OX4 1JF
UK

Blackwell Publishers Inc.
238 Main Street
Cambridge, Massachusetts 02142
USA

British Library Cataloguing in Publication Data
A CIP catalogue record for this book is available from the British Library.

Library of Congress Cataloging-in-Publication Data
Library of Congress data has been applied for

ISBN 0-631-18115-6 (hbk)

Typeset in 10/12 pt Garamond
by Pure Tech India Ltd, Pondicherry
Printed in Great Britain by Hartnolls Limited, Bodmin, Cornwall

This book is printed on acid-free paper

Contents

Preface

This volume is based on essays given at the fourth economic history summer school held at the University of Warwick in 1989. The editors wish to thank Gerardo Marotta, the President of the Italian Institute for Philosophical Studies, Antonio Gargano, the Institute's ever-capable Secretary, and Professor Luigi De Rosa for their support for the original conference and the publication of this series.

Introduction

The nine essays in this volume are concerned with the relationship – or better, the relationships – between agriculture and modern economic growth across a time span that moves from the eighteenth century to the present and in contexts that range from Britain, France and Italy, to Japan, India and Russia. There are, nonetheless, many obvious gaps: it would have been useful to have included studies on other major European economies with distinctive agrarian profiles, such as Germany, Hungary, Denmark or the Netherlands. It would also have been desirable to have included an essay on North America, given the impact of Transatlantic agriculture on all European farming in the second half of the nineteenth century, just as a contribution on China – whose early agricultural history mirrored much of what happened in Japan, even though its subsequent economic destiny may at least initially have had more in common with colonial India – would have been illuminating. But space is finite, and it would in any case be illusory to pretend that a single set of essays should offer a comprehensive empirical survey of the variety of possible roles played by different forms of agriculture in modern economic growth.

The essays that follow do, however, illustrate and develop the ways in which the economic historian's approach to, and understanding of, the relationships between changes in agriculture and the broader process of modern economic growth has undergone important changes in recent years. To some extent, this has been a matter of returning to earlier and unfinished debates that were cut off by a variety of 'stage theories' of industrialization that proved especially influential in the 1960s and 1970s and made an 'agricultural revolution' a prerequisite for subsequent industrialization.

Economists generally ascribe agriculture two sets of functions in economic growth: on one hand, to supply labour, food, raw materials and accumulated capital for industrial or infrastructural investment, while at the same time generating demand, initially in general and for consumer goods but subsequently also for manufactured goods and machinery. As far as supply is concerned, in a pre-industrial economy expansion of non-agricultural sectors can occur only if agriculture is able contemporaneously to release labour for other forms of employment while maintaining or improving existing levels of output (unless, of course, alternative sources of food were miraculously available to support an increased population no longer directly engaged in agriculture). Hence increased productivity is, if not the only, at least the most probable means of avoiding Reverend Malthus's cheerless prediction that increased population would outstrip resources and lead to terrible famine and catastrophe. Agriculture may also be a source of capital accumulation that is subsequently transferred to industry and services through direct or indirect investment, savings or taxation, and may in addition supply an expanding economic system with practical expertise in the form of entrepreneurship or commercial organization, or help to stimulate changes in legal structures, terms of land ownership and tenancy, commercial institutions, the organization of credit and insurance, for example, that may prove of wider application and utility. Agriculture may then in turn generate various forms of demand to sustain economic growth, by providing markets for consumer and manufactured goods or, at a later stage, for an array of machinery, chemicals and other industrial products.

These functions have, however, always proved easier to define in theory than in empirical reality, are extremely difficult to measure, and have in any case to be historically demonstrated. This is not easily done, and the fact that agriculture is assigned both supply-side and demand attributes hints at the complexity of the relationship. But agriculture as a distinct 'sector' remains difficult to measure and to define. Not by chance, questions of measurement and definition recur throughout the essays that follow and are inseparable. In its historical contexts, agriculture is an activity that is not easily isolated either from other branches of economic activity or from prevailing forms of social and political organization that determine how agricultural rents and profits are distributed. Even in more narrow, technical terms the measurement of output and changes in productivity is no less challenging, and this remains true for more contemporary, as well as for more distant, periods. No less problematic – returning to the simple supply and demand model outlined above – are the means by which profits, savings, and capital are, or may have been, transferred from agriculture into other sectors of the economy.

Despite these uncertainties and imprecisions, the notion that agriculture and commerce were the original source of the capital that made industrialization possible was as central to Karl Marx's analysis of the development of industrial capitalism, which for all its exploitative character would in Engels's term 'free men from the idiocy of rural life', as it was for the proponents of free market capitalism. Despite the fact that the notion of an 'agricultural revolution' was always more metaphor than model, it gained added persuasiveness from the example of Britain's first 'industrial revolution'. This was the historical example that inspired W. W. Rostow's triumphal 'stage of growth theory' that set successive 'revolutions' in agriculture, commerce and transportation, but in agriculture above all, as the necessary prerequisites for the qualitative 'take-off into self-sustained growth carried forward and upwards by industrialization.[1]

The metaphor proved highly influential, and not only among economic historians, not least because, in highlighting the critical role of Britain's eighteenth-century agricultural 'revolution', it seemed to provide compelling explanations both for Britain's subsequent and precocious industrialization, and at the same time for the relatively slow pace of continental European emulation. The 'agricultural revolution' provided Britain with precisely those factors of growth (a non-agricultural labour force, capital and a pre-formed consumer market) that her Continental emulators conspicuously lacked. Indeed, across the English Channel the French Revolution's consolidation of small peasant properties could be taken as the principal reason for France's failure to maintain rates of economic growth that in the eighteenth century had paralleled, if not outpaced, her off-shore rival.

Elsewhere in Europe, the persistence of a large, pre-modern peasant sector could also be identified as the main obstacle to spontaneous economic growth. Alexander Gerschenkron extended the assumption that spontaneous, market-led growth was premised on the modernization of agriculture when he claimed that state intervention and the development of investment banks could, in certain historical circumstances (in Russia at the end of the nineteenth century, in Italy and other industrial late-comers, and perhaps partly too in Germany), act as a surrogate for the missing 'agricultural revolution'. At a price, however, because this was also likely to give rise to a different, more dirigiste and dualistic process of economic growth that was led from above rather than driven from below, anticipating the conclusion of those like Barrington Moore Jnr who saw in the survival of a pre-modern agrarian sector the roots of twentieth-century dictatorships and ultra-nationalist politics.[2]

Over the last two decades, these arguments and the assumptions that underpin them have come under increasingly critical scrutiny.[3] Looking back from the end of the twentieth century, neither development nor underdevelopment seem to be permanent conditions, while confidence in the capacity for self-sustained growth has given way to new concerns about the 'limits of growth'. Empirically, the stage theory of growth has proved difficult to apply to historical realities: there has been little consensus in identifying the timing of, never mind measuring, the supposed 'take-off' of the different modernizing economies. Developmental strategies premised on the notion of making agriculture pay for public investment in agriculture had generally proved unsuccessful. Doubts have grown, too, about the validity of adopting any single, unilinear model of modern economic growth – and especially one based on the example of Britain simply because as the 'first' industrializer the British pattern of growth must necessarily hold the only true secrets of spontaneous economic growth.

All this has made economic historians more sensitive to the diversity of different patterns of modern economic growth and more alert to the ways in which economic growth has derived from broad clusters of changes affecting whole economies and societies and often over extended time periods, rather than from a sudden, heroic 'breakthrough' pioneered by a single 'leading sector'. The leaps and discontinuities inherent in the notion of an industrial 'revolution' are displaced by new emphases on the longer term and more capillary processes of growth. In this context of longer-term growth, the issues that move to the forefront of the economic historian's attention are the interactions between different sectors, and also the human agencies of economic change – the choices and adaptations made at different times and in different places by historical economic actors and subjects that make visible the working of the human hand in economic change.

With the demise of the British-based unilinear model of modern economic growth, interest in the role of agriculture in the development of modern industrial economies has, if anything, grown. This has also reopened the debate on agriculture's contribution to British industrial growth, the terms and significance of which are reviewed in Mark Overton's opening essay in this volume.[4] While it is widely accepted that British agriculture had achieved levels of productivity that were in comparative terms quite exceptional by the eighteenth century, and indeed earlier, and that this constituted one of the central 'peculiarities' of the British pattern of economic growth, measuring changes in productivity and explaining why they occurred remains much more problematic. While acknowledging the importance of long-term trends and developments in the increase in agricultural productivity, Overton argues that the term 'agricultural revolution' is still a valid description of the unprecedented increases in output, land productivity and labour productivity that occurred after 1700 'marking a decisive break with the past'. Working from new estimates of output, land productivity and land area, Overton challenges the idea that increases in agricultural productivity developed over long periods, but he also argues that accumulated improvements in knowledge, management and organization were probably more decisive than science or technology in enabling British farmers to raise output in the eighteenth century without damaging the ecological balance through the adoption of biological technology (the four-course rotation) and through raising labour productivity.

As Overton notes in his conclusions, the levels of improved productivity achieved in Britain in the first half of the eighteenth century would be outstripped in the first half of the nineteenth century, the period which Michael Thompson has christened the 'second agricultural revolution'. The precocious marginalization of the primary sector was certainly central to the 'peculiarity' of Britain's pattern of industrialization and economic growth, and, in the course of the nineteenth century, this peculiarity became even more pronounced. This is evident from Michael Thompson's essay that reconsiders the supposed demise of British agriculture in the years 1870 to 1914. Denied the protective tariffs of its European neighbours, British agriculture is generally thought to have become increasingly uncompetitive in these years when more adventurous free-trade rivals, like Denmark and the Low Countries, began to develop new specialist agricultural industries that made heavy inroads into British markets. 'To surrender to American, Canadian, Australian, Argentinian, Russian and Indian wheat was one thing. To surrender to Danish bacon or Dutch butter and cheese was another. . . .' But Thompson goes on to argue that appearances were deceptive. The development of a major new market in producing and distributing fresh milk revealed the ability of British farmers to seize comparative advantages, because this was a commodity with which overseas dairy producers (the Dutch, Danes and the Irish) could not at this time compete. British farmers also retained a strong share of the fresh-meat markets, despite overwhelming competition from the United States and from South America for chilled and frozen meat. Above all, this period witnessed a significant 'industrialization' of agriculture as British farmers between 1870 and 1914 invested more heavily and extensively in machinery and fertilizers than European competitors. In both cases, the principal reason was to reduce labour costs (previously the largest item in production costs). Expansion in livestock production also created a growing demand for industrial feeds such as oilcake which, together with demand for fertilizers, made agriculture an increasingly important consumer of industrial products and 'pointed the way towards the later twentieth-century model of industrialized agriculture'. Reduced production costs, combined with a fall in agricultural rents, enabled farmers' incomes to rise in relative terms despite falling prices.

Thompson concludes that, although British agriculture did not constitute a major source of industrial demand before 1914, it had shown a remarkable capacity to adapt: 'At best, a slimmer agriculture in 1914, really was fitter, economically, than it had been in 1870: British consumers enjoyed cheap food, the cheapest in Europe, and British farmers and farm workers enjoyed higher real incomes'. As well as underlining the continuing 'peculiarity' of agriculture in the British economy in this period, Thompson's essay demonstrates the complex interplays and interdependencies that characterized the relationship between agriculture and industry as Britain moved towards a mature industrial economy. It also illustrates the difficulties (already addressed in Mark Overton's essay) of measuring reliably real rates of productivity and their changes, and emphasizes the importance of regional differences and varied institutional settings. But, within those contexts, he also shows how changes and adaptations in British agriculture in these years resulted from the choices, options and calculations made by individual landowners, tenant farmers, small cultivators, labourers and farm hands as they responded to changing conditions and opportunities.

Government policy, on the other hand, seems to have played a slighter role even though the decision to adhere to free trade before 1914 effectively abandoned British agriculture to its own devices on the grounds that the British Navy would always be able to ensure that Britain was fed. The debates on the impact of government policy, as opposed to more spontaneous forces of growth, change and adaptation within agriculture itself, are given a more contemporary focus in B. A. Holderness's essay which examines the remarkable performance of British agriculture in the period from the end of World War II to the oil crisis of the early 1970s, which preceded Britain's entry into the European Common Market and the Common Agricultural Policy (CAP).

Holderness describes these decades as a 'Third British Agricultural Revolution'. In the 40 years before 1975 the gross output of British agriculture trebled, and self-sufficiency in temperate products rose from 40 per cent to 60 per cent, with huge increases in labour productivity (the agricultural work-force was halved in size, while horsepower involved in farming rose some tenfold). This achievement is to some extent hidden by the fact that British agriculture was already more marginal in generating Gross National Product than any of its European competitors, which meant that new advances in productivity could not reverse this marginalization. The advances were also achieved primarily through 'the envelopment of agriculture with science and technology', with the result, as Holderness argues, that it is no easier in the contemporary than in earlier periods to isolate agriculture as a discrete sector with easily identifiable performance indicators. These tendencies were intensified as the mass distribution of frozen and packaged foodstuffs generated major agro-industries.

Holderness also sets these developments in comparative context when he challenges the claim that the success of post-war British agriculture has been due mainly to government policies and subsidies. While emphasizing again the difficulty of measuring increases in labour and land productivity,

Holderness claims that advances were high and derived primarily from the 'interpenetration of industry and agriculture'. But high increases in productivity in the post-war period were not unique to Britain, and 'equal or superior gains accrued in some [European] rural economies dominated by much smaller farms ...', '... the British experience after the war was not unique. The third agricultural revolution encompassed all the temperate latitudes outside the Soviet bloc. It took place in spite of institutional arrangements, under Protection, Deficiency Payments and Free Trade, in inward-looking and in world-centred agrarian economies.' The particular characteristic of British agriculture within this more general 'Third Agricultural Revolution' was, therefore, that it became the most highly mechanized in the world. Holderness argues that this was a response to the shortages of rural labour in Britain rather than a cause of rural depopulation.

This had little to do with subsidies, Holderness argues, although in the mid-1950s these were higher than in any other European state except Switzerland, and significantly higher than in France, Germany and Italy. But subsidies did raise net farm incomes and, by creating a sense of confidence among farmers, encouraged unprecedented land purchases that caused an 'inexorable increase in the proportion of agricultural land in this island owned by the occupiers and a massive increase in mortgage debts and hence interest payments borne by agriculture'.

The three essays that follow move back to consider Britain's nineteenth-century European neighbours and competitors – France and Italy. Roger Price examines the reasons why the development of new railway networks in France in the nineteenth century did not always provide the incentives for agriculture that might have been expected. In contrast with Britain even a century earlier, France's greater size, dispersed population and lack of natural communications constituted a significant comparative disadvantage for agriculture and meant that 'the shortcomings of the communications network overwhelmingly oriented actors in the market place towards the local and regional, and that distinctive marketing regions remained only loosely connected'.

The creation of a railway network might therefore have been expected to have an especially dramatic impact on agricultural output and specialization, especially because the urban population in France increased (as a percentage of total population) from 25.5 per cent in 1851 to 44.2 per cent in 1911. But Price argues that the creation of an integrated national agricultural market proved slower and more partial than liberal economists and administrators had anticipated. The impact of the railways on agriculture was highly differentiated, not least because the networks excluded many agricultural regions, especially in the west and south-west, from urban markets, while differential freight tariffs benefited some destinations (especially Paris) at the expense of others. The railways also opened French markets to import penetration in ways that deprived many French regions of their former markets. A a result, Price argues, French rural communities were hesitant in the face of changes that seemed to bring greater uncertainty rather than clear new opportunities, and these apprehensions were increased by the combined effect of improved communications and imports in lowering the price of agricultural products. The result was increasing opposition to free trade in rural France and the introduction after 1881 of new tariffs to protect livestock and French cereals. Protectionism and government pressure on the railway companies to revise freight rates led to an increase in inter-regional agricultural trade, but Price argues that this occurred in conditions that did not encourage farmers to innovate. While acknowledging that free trade and the railway network brought about profound and irreversible changes in the organization of agricultural production and markets in nineteenth-century France, Price also emphasizes the relatively slow and regionally differentiated ways in which these changes occurred. Specialization occurred in some regions but not all, and he concludes that 'only a minority of efficient cultivators were primarily oriented towards the market . . . for others, price depression and tariff protection reduced incentives to sell on the market and the pressure to innovate. The potential for change was reduced and its progress notably slowed.'

Price's conclusions tend to confirm the view that the development of French agriculture in the nineteenth century was slower than it should have been. This is a view that is partly shared by Colin Heywood, who nonetheless sets these conclusions in a different context and argues that relatively slow rates of growth in agriculture and the survival of an extensive small peasant farm sector did not function as a brake on the expansion of France's industrial economy in this period. Indeed, he turns that argument on its head to argue that the slow development of French industry in the last quarter of the nineteenth century denied French farmers and agriculture opportunities for expanding output and productivity. Heywood steers a middle passage between those who argue that French economic growth was among the slowest in Europe in the final quarter of the nineteenth century and more recent claims that, when aggregate figures are related to France's slower rates of demographic growth, the overall per-capita growth rates were only slightly behind Germany in this period and ahead of Belgium, the United Kingdom and Italy. But Heywood argues that, despite the catastrophic impact of phylloxera on wine production and of the 'Great Depression', the strongest sector of the French economy in these years may well have been agriculture and he challenges the notion that Third Republic France was a *société bloqué*, in which a conservative middle class allied with a traditionalist peasantry to slow down the pace of modernization. In agriculture as in industry, Heywood argues, small-scale enterprise was the norm in the French economy, and was neither inefficient nor unproductive. Small peasant farms were particularly well suited to engage in a range of different agricultural, commercial and industrial activities (*pluriactivités*) in ways that enabled them to adapt with great flexibility to changing opportunities and conditions. He challenges the idea that the main demand for agricultural protectionism came from peasant farmers: tariff protection was adopted relatively slowly in France, and the pressure came first from industrial interests and from the large-scale northern farmers. The peasant farmers, whose capacity to adapt and their preference for quality rather than mass-market products meant that they were less affected by foreign competition, had benefited from free trade and often stood to lose from protectionism.

Warning against trading one unqualified appraisal of the impact of protectionism for another, Heywood notes that there is 'an air of unreality' in the debate because French governments had little alternative but to follow where Germany, Austria-Hungary and Italy had already led. Only Britain, with its relatively small agricultural population could withstand the political pressures for protectionism, and, according to Paul Bairoch's calculations, all the countries that adopted protectionism in this period experienced faster growth than Great Britain.[5] There is no evidence, Heywood argues, that the persistence of small peasant farms starved French industry of labour in this period, whereas the predominance of highly skilled artisan labour organized in small workshops meant that French industry did not require a massive influx of labour from agriculture. But although there did not appear to be any shortage of capital for industrial development, he acknowledges that France was slow to develop agricultural credit and to provide technical training comparable with Denmark and Germany. This did not mean success or failure judged against some abstract model of maximum utilization of resources or against the British model of industrialization with its dark satanic mills and 'Coketowns'. What is does suggest, Heywood concludes, is an alternative pattern of growth marked by a specifically French 'culture of production' that favoured small enterprise in industry and agriculture, and the capacity to respond and adapt in ways that avoided dramatic surges while sustaining steady increases in production and wealth.

Paul Corner's essay on Italy follows a similar line, and starts by comparing the ways in which France and Italy have conventionally been partnered as examples of slow or unsuccessful industrializers. Deriving from a schematic model that identifies modern economic growth exclusively with the development of large-scale high-technology industries and the rapid shrinkage of the primary sector, Corner argues that this perspective fails to take adequate account of broader developments that, over the longer term, have created economic growth in Italy. Focusing on the role of peasant farms and peasant families in economic growth, he argues that, in Italy as in France, agriculture may have played a more substantial and significant contribution than is recognized. This, Corner warns, is not to argue that peasants were highly efficient producers, but rather that 'traditional agriculture played a very different role in the process of industrialization from that (passive, negative role) usually ascribed to it'. That role cannot be understood solely in terms of capital accumulation or the productivity of labour because it was determined above all by the ways in which peasant families responded to economic change.

The protagonists of change were the peasant-worker families, and like Heywood, Corner emphasizes the capacity of peasant farms and families to engage in a range of 'pluri-activities'. His principal examples are drawn from the northern Italian peasant families who, from the early nineteenth century, were engaged in raising silkworms on their land, and then became involved in a variety of processing, manufacturing and commercial activities. He emphasizes that agricultural and manufacturing activities were combined for reasons that were as much cultural as economic. Patriarchal peasant families retained a strong attachment to the land because it alone offered some degree of autonomy. But the way to retain land lay in developing new forms of by-employment for unmarried sons and, above all, for daughters. As in much of France, urban working conditions in late nineteenth-century Italy were unattractive for rural families, and this strengthened the peasants' determination to preserve the peasant-worker household. The result was the formation of a work-force in Italy with features that differed from the 'classic' paradigm: it was 'non-urban, non-proletarianized and not predominantly male . . . extremely flexible, extremely attentive to economic opportunity, given to high levels of self-exploitation, not without some experience in manufacturing, and prepared to use initiative and take risks in order to achieve economic independence'.

While this offers an explanation for the origins of the remarkable tradition of family enterprise that has characterized contemporary economic development in much of central Italy,[6] Corner argues that family adaptation and rural pluri-activity has contributed much more broadly to Italy's characteristically 'wave-like' process of economic growth. In part this can be seen as a way of responding to Italy's relatively vulnerable position in a developing international economy (family based pluri-activity offering a range of opportunities to spread risk and retain economic alternatives). He also argues that the regional characteristics of Italy's long-term economic development reinforce the decisive importance of family based economic strategies, because it is in precisely those parts of Italy where peasant families have been least cohesive and have enjoyed least economic independence (above all, in the south) that economic growth has been slowest and most uncertain.

Like Heywood, Corner emphasizes the long-term characteristics of economic growth and also the determining role of social and cultural forces. These themes are equally prominent in Kaoru Sugihara's essay on the role of agriculture in Japanese industrialization. He, too, sets out by challenging the appropriateness of applying a European (or British) model of economic development to a country like Japan in which the social structure and resource mix were clearly quite different. This model has led historians to describe Japan's industrial development in the nineteenth and early twentieth centuries as 'lopsided': peasant agriculture provided a reserve of cheap labour for rapid industrialization that resulted mainly from government intervention and direction. The backwardness of peasant agriculture inhibited the development of a domestic consumer market, however, and brought about a crisis in the 1930s which Japanese industry attempted to escape through imperialism and aggressive overseas expansion.

Echoing arguments raised by Heywood and Corner in the context of France and Italy, Sugihara argues that this model is misleading because it overstates the importance of government intervention and neglects the contribution of agriculture and the peasant economy to Japanese economic growth over the longer term: 'The Japanese experience might be better assessed if we abandon this assumption of a unilinear (and often Eurocentric) pattern of development'.

From well before the time of the Meiji Restoration (1868), Japanese agriculture was highly productive, specializing in the production of rice, cotton and silk while maintaining a favourable balance between land and population. This was achieved primarily through the intervention of two institutions: the village community and the peasant household. As in Italy and in many parts of France, Japanese industrial development in the eighteenth century was almost exclusively rural (accompanied – as in many parts of Europe – by the decline of older urban manufacturers and trades) and offered the peasant family additional means of employment. Following the Meiji Restoration, exports of rice, raw silk and tea paid for Japan's capital imports, while agriculture also provided the Japanese state with its principal tax revenues. While attempts to introduce Western style agricultural changes proved unsuccessful, increases in agricultural productivity resulted, as in earlier periods, from the adoption of labour intensive technologies. The greater part of Japanese industry (which grew from 15 per cent to 32 per cent of NDP between 1885 and 1915) was rural. Cotton and silk-reeling were the most important branches, so that the 'technological development of small scale agriculture and that of traditional industry reinforced each other . . .'

Integration with agriculture and peasant family economies provided the textile industry with low cost labour, while the persistence of traditional values provided industry with a highly obedient and disciplined labour force (as in many Italian industries, but unlike India).[7] Cheap rural labour enabled Japanese manufacturers to invest in, and use, western technology, and did not imply any lack of competitiveness. Throughout the inter-war period Japanese industry retained its rural characteristics and 'the rural household remained the only major societal base under which industrial labour was produced'. This pattern was not broken until after World War II, when Japan experienced unprecedented urbanization and the displacement for the first time of small enterprise by big business. Sugihara concludes that the introduction of lifetime employment, seniority wages, welfare facilities, occupational pensions and 'companyism' as an ideology in Japanese industry after World War II can be best understood as a strategy to create a new stable urban household system to take the place and play the role of the former rural household that, together with the village community, had played the key role in shaping Japan's earlier economic growth.[8]

Many of these themes recur, albeit in different form and context, in David Washbrook's essay on agriculture and industrialization in colonial India. Taking a long-term view of the stagnation of 'one of the early modern world's leading manufacturing and merchandizing economies', Washbrook points out that the economic history of colonial India contradicts the premises of the 'stages of growth' model of industrialization at various turns. Between 1820 and 1850, India's manufacturing sector was 'de-industrialized' at the same time that output from agriculture was increasing, whereas the most marked period of industrial expansion (between the two world wars) occurred at a time when agriculture was wracked by the world Depression. Punjab was 'the one unequivocal "success" story of colonial agriculture but generated little industry, whereas Calcutta became one of India's principal industrial metropolises despite the fact that West Bengal's agrarian economy had been in decline since the 1860s'.

Washbrook makes the political economy of British colonial administration the context for discussing Indian agriculture and its different functions. Far from failing to generate surpluses, Indian farmers were the principal source of revenues for British colonial administration, indigenous aristocracies and intermediaries. Agriculture, in short, generated surpluses that were devoted mainly to purposes other than industrial investment.[9] India also produced huge quantities of industrial raw materials, most of which were exported to other parts of the world. If anything, Indian agriculture released too much, rather than too little, labour in ways that discouraged technological innovation, while, unlike the Japanese family based rural labour force, the casualized Indian labourers formed a 'pauperized' peasantry that made industrial unrest and protest prolonged and endemic.

British power destroyed the internal demand generated by India's princely courts and administrations, while British rulers and a new Indian 'protobourgeoisie' conspired together to siphon off surpluses from agriculture through taxation and a 'ferocious "rental offensive" '. But colonial rule and contact with the world of modern capitalism 'actually caused several of the worst problems' for Indian agriculture. Washbrook shows how the subcontinent provides another example of a railway system that was 'a distinctly mixed blessing for the futures of both industry and agriculture'. Careless engineering strategies turned rice-lands into malarial swamps and caused massive deforestation, and above all created endemic dearth because crops moved towards areas of highest prices. These were nearly always those struck by famine, thereby causing 'dearth-level prices' to spread throughout India. Dearth in turn reduced production, while attempts to introduce more scientific forms of farming and modern irrigation systems proved inept and damaging. Peasant entrepreneurship in certain western zones (Punjab, Gujerat, interior Tamilnadu) won through against all these odds, but the main obstacles to agricultural development in India are identified in colonial rule, which diverted revenues from agriculture to maintain a massive military machine and world empire, and modern industry which utilized 'an extensive range of raw materials produced by the Indian peasant' while giving back very little, above all because an impoverished peasantry provided a rich source for industrial capital to exploit. Hence Washbrook concludes that 'while Indian agriculture may have "failed" the mass of the Indian population, and especially the peasantry, the nature of its development succeeded for the British empire and for both metropolitan and indigenous Indian capital'.

As well as illuminating the exploitative character of colonial administration, Washbrook's analysis also underlines the critical importance of the impoverishment of the peasant family economy whose contribution to the development of industry and agriculture in Japan and in many European countries is the subject of essays discussed above. The peasantry have also been central and tragic protagonists of the history of Russian and Soviet agriculture in the twentieth century, and in the final essay in this volume Mark Harrison provides further evidence of the critical problems posed by the absence of a stable and independent agricultural community – something that the Soviet *kolkhoz* attempted to replicate, without succeeding.

Harrison's discussion of Russian agriculture develops an extended critique of the notion that agriculture is the mainstay of industrialization. Showing how the idea that agriculture could pay for public-sector industrial development ran through Russian thought and administrative policy from the tsars to the Bolsheviks, he demonstrates that this was never successfully applied in Russia and was never shown to work. 'Neither before nor after the revolution has there been demonstrated any direct link from forced saving of the peasantry to industrial formation.'

Before the revolution, the main obstacle to agricultural development was taxation which (as in British India) was devoted to administrative and military expenditure, not industrial investment. In the 1920s, however, Soviet governments were willing to commit 'a really significant share of budget revenues to paying for industrial capital formation'. But these attempts foundered on the inability of Soviet agriculture to yield adequate surpluses. Although a late convert to the idea of making agriculture pay for industrial investment, Stalin was nonetheless ready to adopt drastically coercive measures to achieve this end, which, starting in July 1928, included the introduction of new procurement systems, the physical liquidation of the peasantry and the collectivization of farming. But despite the appalling human costs, these measures failed to increase the 'tribute' from agriculture: procurements caused famine, and collectivization caused production to collapse. The Soviet planners had no clear idea of how the collective farms (*kolkhoz*) should be structured or organized or how they should operate. As a result, the collectivized sector staggered through *ad hoc* experiments that relied primarily on repression and coercion.

Despite some expansion of the private sector during the war, there was no significant change until after Stalin's death when first Krushchev and then Brezhnev gave agriculture increasing priority. Efforts were made to improve the management standards of the *kolkhoz* and to improve the conditions of *kolkhoz* workers. Food output improved and so did the diet of Soviet citizens, but nonetheless Soviet agriculture proved unable to meet growing demand and, in the 1970s, the Soviet Union was forced to rely on large-scale imports of wheat and other foodstuffs. Some progress was achieved, and in the 1950s and 1960s, the growth in output was higher than in the United States. But this did not narrow the efficiency gap, which widened again in the 1970s. Harrison argues that the failure of Soviet agriculture to meet the targets set for it in the post-war period can be explained only partly in terms of poor resource management, lack of worker incentives, irrational procurement prices, rapid policy changes and over-centralized controls. Despite these problems, 'the Stalinist legacy was overcome, and agricultural problems more and more began to resemble those of the economy as a whole'. This was why Gorbachev's attempts to reorganize the *kolkhoz* system and the organization of agriculture proved impossible to implement without a wider reform of the Soviet economic – and political – system. 'Rural producers would have to make the difficult transition from passively surrending food surpluses and receiving supplies to the responsible exercise of power, with equal citizen's rights of participation in, and control over, the fate of the rural society and ecology'.

Mark Harrison's conclusions echo many of the themes raised in previous essays: agriculture's reluctance to conform to the performance paths preset by economic theory; the tendency for policies designed to channel agricultural development in one direction or another, or to encourage transfers and linkages to and from other sectors, to have consequences quite different from, and even in contradiction to, those intended. The shortcomings of economic policy with regard to agriculture reflect the inadequacies of the theoretical formulations that have guided and shaped policies. These in turn suggest an inadequate understanding of the variety and complexity of the roles that agriculture may play, and historically has played, in modern economic growth. One of the most fundamental difficulties lies in the attempt to treat agriculture as a discrete and self-sufficient economic sector, with behaviour that can be determined and influenced in isolation from the wider economic, political and social systems of which it is a part. The progressive intermixing of industrialization, science, technology and farming has made this integration even more complex, but the essays in this volume dealing with the earlier periods also indicate that even then a 'single-sector' approach to agriculture serves primarily to illustrate that the process of change and economic growth in agriculture cannot be understood as independent variables because they have at every point been related to, and impacted on, much wider and more complex networks of economic, social and institutional relations and structures. That is why economic historians have begun to pay closer attention to the variety of different forms that changes in agriculture have taken in different historical contexts and periods, and, in doing so, have also taken increasing account of the role played by those most directly involved in farming, whose perceptions of advantage and risk emerge from all of the essays in this volume as the most persistent and critical agents and determinants of change.

Notes

1 W. W. Rostow, *The Stages of Economic Growth: a non-Communist manifesto* (Cambridge, 1960).
2 Gerschekron, *Economic Backwardness in Historical Perspective* (Cambridge, 1962). Barrington Moore Jnr, *Social Origins of Dictatorship and Democracy: Lord and Peasant in the Making of the Modern World* (Boston, 1966).
3 For recent surveys *see* the first volume in this series: Peter Mathias and John A. Davis, *The First Industrial Revolutions* (Oxford, 1989); E. L. Jones (ed.), *Agriculture and the Industrial Revolution* (Oxford, 1974); id. 'Agriculture 1700–80' in R. Floud and D. McCloskey (eds), *The Economic History of Britain since 1700*, Vol. 1, *1700–1860*, pp. 66–87 (Cambridge, 1981); J. Beckett, *The Agricultural Revolution* (Oxford, 1990); and the introductions to P. K. O'Brien and R. Quinalut (eds), *The Industrial Revolution and British*

Society (Cambridge, 1993) and F. M. L. Thompson (ed.), *Landowners, Capitalists and Entrepreneurs: Essays for Sir John Habbakuk* (Clarendon Press, Oxford, 1994).

4 *See* P. Mathias, 'Agriculture and Industrialization' in Mathias and Davis (1989).

5 *See* F. M. L. Thompson's essay in this volume for a different view of the performance of British agriculture in these years.

6 *See* Davis in: Peter Mathias and John A. Davis (eds), *Enterprise and Labour from the Eighteenth Century to the Present* (Oxford, 1996).

7 For Italy, *see* Corner below pp. 129–47; for India *see* Washbrook p. 167–91.

8 *See also* T. Matsamuro in Mathias and Davis (1996).

9 *See also* Mark Harrison's comments on agriculture and taxation in Tsarist Russia pp. 192–207.

1

Land and Labour Productivity in English Agriculture, 1650–1850

Mark Overton

Introduction: 'Agricultural Revolutions'

For many years the fortunes of English agriculture from the early modern period onwards have been discussed in terms of the existence or otherwise of particular 'agricultural revolutions'. At least five periods of 'revolutionary' change have been identified between 1560 and 1880, and each has been characterized by a different combination of 'significant' agricultural developments. Debate about the character and chronology of these 'revolutions' has reached something of an impasse in recent years, partly because they have not been backed up by sufficient quantitative evidence, and partly because too little attention has been given to the criteria by which the 'significance' or otherwise of agricultural change should be judged.[1] Yet the substantive issues with which the debate has been concerned remain of central importance to understanding the development of English agriculture and of the English economy. This chapter will attempt a reassessment of the development of English agriculture from the seventeenth to the nineteenth centuries by recasting the debate on the 'agricultural revolution' in terms of changes in output and productivity. It will argue that the concept of agricultural productivity provides a more useful yardstick by which to gauge progress than vague and inconsistent notions of 'revolution', and re-evaluate agricultural developments in the period in the light of some quantitative estimates of productivity and output.

Writing more than a century ago, historians were clear that the agricultural revolution took place in tandem with the industrial revolution in the century after about 1750. It was characterized by a series of technological changes – new crops, new crop rotations, new breeds of livestock, and new implements and machines – and by a series of changes in the institutional structures of farming, the most important of which was the transformation of property rights through the process of parliamentary enclosure. The major technological innovations emphasized in this account were two fodder crops, turnips and clover, which expanded livestock carrying capacity and therefore supplies of manure – the main fertilizer of arable land. This raised soil fertility and hence yields. The new crops were grown in a rotation in which grain crops alternated with fodder crops, replacing the old rotations in which several grain crops were taken in succession followed by a bare fallow. The ultimate expression of these principles was in the Norfolk four-course rotation of wheat, turnips, barley and clover. Other eighteenth-century improvements included the selective breeding of livestock, which changed the size and shape of animals, but more importantly improved the rate at which food was converted into meat.

According to this view of the 'agricultural revolution', the prime achievement of the period lay in feeding a rapidly growing population, so that the agricultural sector of the economy did not function as a break on industrial expansion and urbanization. Although, as we shall see, this view has been challenged, it still remains persuasive. The English population grew by 11 million from 1750 to 1850 and, although food imports increased substantially, it has been estimated that 6.5 million extra mouths were being fed by English agriculture in 1850 compared with 1750.

Nevertheless, several authors think that earlier developments were of more significance. Population also doubled between about 1550 and 1750, and the extra food required to feed that growing population was also provided by English agriculture. In its most extreme form the argument for an 'agricultural revolution' during the early modern period dismisses the significance of agricultural changes after 1750 as irrelevant, and maintains that some technological innovations occurred much earlier, including the introduction of fodder crops, new crop rotations, and field drainage. Emphasis is also placed on convertible or ley husbandry, in which the distinction between permanent grass and permanent tillage was broken, and grass was rotated round the farm.

Less extreme versions of this view hold that rapid technological change in the form of cropping innovations took place in the century after 1650, although some authors have described these innovations as amounting to an 'agricultural revolution'. In contrast to the following century, population growth remained roughly static after 1650, so that the importance of the period lies in a rapid 'transformation in techniques'. These led to increases in grain output per acre, and a rise in total output evidenced by rising grain exports. The processes by which output was increased are virtually the same as for the post-1750 revolution; a rise in the fertility of the soil through the introduction of turnips and clover and their associated crop rotations. The stimulus for change is seen as a run of sluggish grain prices which squeezed farmers' profits. This caused them to keep more livestock, and more importantly, to lower unit costs of production by raising yields through the innovation of fodder crops.

The most recent verdicts on agrarian developments from the sixteenth century can be found in the Cambridge *Agrarian Histories*.[2] For the century after 1650, Volume V concludes that a depression in grain prices prompted innovation and enterprise, but that the full harvest of this ingenuity in the form of an 'agricultural revolution' was not to be reaped until after 1750. On the other hand, the succeeding Volume VI, dealing with the period 1750–1850, regards the agricultural changes of this period as unworthy of a 'revolutionary' label and regards them merely as a limited preparation for the greater changes yet to come in the twentieth century.

Although the criteria for these various 'revolutions' vary, all emphasize the significance of technological changes, particularly those relating to the innovation of new crops. Sometimes explicitly, but more often by implication, these are seen as important because they improved 'productivity', usually taken as grain yields per acre, which subsequently led to increased output. Yet concepts of agricultural productivity remain woolly and ill-defined: the productivity of land is usually equated with grain yields per acre (usually for wheat alone) while discussions of the productivity of labour have been subsumed in the issue of the 'release' of labour from the agricultural to the industrial sector during the industrial revolution.

Agricultural Productivity

Productivity can be simply defined as the ratio of output to input. In practice, productivity indices vary considerably, depending on the combinations of outputs and inputs that are considered and the units in which they are measured, but the two productivities usually considered as most significant to the study of agriculture are those of land and labour.[3] The relationships between output, land productivity, and land area can be illustrated symbolically as follows.

$$Q = QCH + QA \quad (1)$$
$$QC = TC \times YC \quad (2)$$
$$QCH = QC - S - CF \quad (3)$$
$$TAR = TC + TF + TFAL \quad (4)$$
$$QA = A \times YA \quad (5)$$

$$A = (TPG \times YPG) + (TF \times YF) + CF \quad (6)$$
$$T = TAR + TPG \quad (7)$$

Agricultural output (Q) is defined as the output of cereal crops for human consumption (QCH) plus the output of animal products (QA). Cereal crop output (QC) consists of the area under cereals (TC) multiplied by the yield per unit sown (YC). The quantity available for human consumption is less than this because seed (S) must be retained for the following year's crops, and some cereals are fed to animals (CF). Given the state of agricultural technology before the nineteenth century, the total arable area (TAR) must also include fodder crops (TF, such as peas, beans, turnips and clover), an area of fallow ($TFAL$), or both.

The output of animals (QA) is given by the number of animals (A) multiplied by the output of animal products (meat, dairy products, tallow, wool, hides and skins) per beast (YA). This is partly a function of how much they are fed, but it also depends on the rate at which animals convert food into these products. For example, the output of meat depends on the number of animals, the quantity of food they receive, the rate at which they convert food into meat, and how long they take to 'finish' (in other words, be ready for the butcher). The number of animals is related to the food available for them, which is measured by the area of permanent grass (TPG) multiplied by the yield of grass (YPG) plus the area of fodder crops in arable rotations (TF) times their yield (YF), and the quality of cereals fed to them (CF).

It should be evident from these equations that there are many possible measures of land productivity. The most straightforward is simply the total output of agricultural products divided by the agricultural area (Q/T). The index of land productivity more commonly used by historians employing farm-based data relates the output of a particular crop (usually wheat) to the area on which it was grown (QC/TC); this is usually calculated in terms of the volume of grain per unit of land sown (Y/C), as bushels per acre. Yields calculated in this way, however, may not be a reliable guide to the overall course of land productivity or to output. Yields are directly related to the area of fallow and fodder, so a farm with a relatively large fallow area might have higher yields per sown acre (QC/TC) but much lower yields per unit of arable (QC/TAR) than a comparable farm with a lower fallow proportion. Obviously grain yields reveal nothing about the course of livestock productivity. This can be measured by relating livestock output to the number of animals (QA/A), or to the area supporting the animals as shown in Equation 6.

Just as land productivity is calculated by dividing output by land area (Q/T) so labour productivity can be calculated by dividing output by the number of agricultural workers. This index does not account for the length of time those employed in agriculture actually spend working, however, so a complementary measure of labour productivity is derived by dividing agricultural output by the number of worker-hours per annum. The calculation can be further refined to take account of the respective contributions of men, women, children, and seasonal and part-time workers.

The calculation of partial indices of agricultural productivity, be they for land, labour, or some other input, are not in themselves an adequate guide to the overall level of the efficiency of agricultural production. An increase in yield per unit area, for example, could arise at the cost of a decline in one or more of the other factor productivities. Some overall measure which embraces all of the separate factor productivities is therefore required. The measure favoured by economists is total factor productivity, which relates output to a weighted combination of inputs. Total factor productivity is then defined as the residual productivity increase that cannot be attributed to the recorded increase in land, labour, and capital as factors of production. The important point is that changes in individual factor productivities need to be interpreted in the light of changes in other factor productivities when assessing the efficiency of the agricultural sector as a whole.

The importance of productivity measures to agricultural change

The two achievements of the agrarian sector from *c.* 1650 to *c.* 1850 which are commonly regarded as of most significance are an increase in output which was sufficient to break the 'Malthusian trap' and allow population to expand beyond the pre-industrial ceiling, and a 'release' of labour from the agricultural to the industrial sector of the economy which was a necessary precondition for industrialization. Malthus's argument was that the supply of agricultural products was limited by the area cultivated. Once all available land is cultivated (by expanding T in Equation 7) then an output ceiling is set which limits the size of the population. If land productivity can be increased, however, then more people can be fed from the same area of land. By definition, an industrial revolution takes place when a growing *proportion* of the work-force is engaged in industrial, or at least non-agricultural, occupations. For this to happen a smaller proportion of the population must be engaged in agriculture; in other words there must be a rise in agricultural labour productivity.

The crucial contribution of English agriculture in the centuries before about 1850 was to raise land and labour productivity together. Not only was the Malthusian trap avoided and population growth largely unchecked after about 1750, but a smaller proportion of the population was able to produce the extra food, thus permitting an increase in employment in the industrial and service sectors of the economy. Although the general course of these changes is well known, neither the magnitude and timing of output and productivity changes, nor the relationships between them have been charted in any detail. The remainder of this chapter will attempt to do this by using some existing estimates of output and productivity and by producing some new estimates.

Estimates of output and productivity

While it is a fairly straightforward matter to recast the 'agricultural revolution' conceptually in terms of output and productivity, it is quite another to attempt a quantification of those trends from the seventeenth century onwards. Comprehensive national agricultural statistics were not compiled until 1866, and official statistics of grain yields were not produced until 1884. This dearth of reliable data has not prevented historians from attempting to estimate output and productivity from the late seventeenth century onwards, using a mix of contemporary estimates, and ingenious manipulations of proxy data.

1 Output estimates

Agricultural output can be estimated in three ways: by using the size of the population as an indicator for the amount of food consumed; through the use of equations specifying the demand for agricultural products using population, prices, real wages, and demand elasticities; and by making direct estimates of the volume of output based on contemporary opinions.

Table 1.1(a) shows the growth of the English population from the early seventeenth century, but at least two assumptions must hold if the growth

Table 1.1 Estimates of agricultural output
(a) 'Population' method

Years	*Population (millions)*	*Net imports (%)*	*'Output' index*
1601	4.11	0	80
1651	5.23	0	101
1661	5.14	−1	100
1701	5.06	−2	100
1740	5.58	−5	114
1751	5.77	−8	121
1760	6.15	−4	124
1780	7.04	0	136
1790	7.74	+2	147
1801	8.66	+5	159
1831	13.28	+12	226
1851	16.74	+16	272

(b) Crafts and Jackson (percentage change per annum)

	Crafts	*Jackson*
1660–1740		0.61
1700–1760	0.60	
1740–1790		0.04
1760–1780	0.13	
1760–1800	0.44	
1780–1800	0.75	
1800–1831	1.18	

(c) 'Volume' method (value of total agricultural in £m at 1850 prices)

	Value	*Index*
1700	40.12	100
1750	51.11	127
1800	76.46	191
1850	114.46	285

Sources: (a) Population figures from E. A. Wrigley and R. S. Schofield, *The Population History of England, 1541–1871: a Reconstruction* (London, 1981); import and export estimates from a variety of sources: (b) N. F. R. Crafts, 'British economic growth 1700–1831: a review of the evidence', *Economic History Review*, second series, 36 (1983), pp. 177–99; R. V. Jackson, 'Growth and deceleration in English agriculture 1660–1790', *Economic History Review*, second series, 38 (1985), pp. 333–51: (c) The combined value of the output of wheat, rye, barley, oats, potatoes, mutton, beef, pork, milk, butter, cheese, wool, tallow and hides based on the estimates in J. C. Chartres, 'The marketing of agricultural produce', in: J. Thirsk, (ed.), *The Agrarian History of England and Wales VII, 1640–1750: Agrarian Change* (Cambridge, 1985), pp. 406–502; Gregory King's estimates in: J. Thirsk and J. P. Cooper (eds), *Seventeenth Century Economic Documents* (Oxford, 1972), pp. 782–3; and B. A. Holderness, 'Prices, productivity and output', in: G. E. Mingay (ed.), *The Agrarian History of England and Wales, VI, 1750–1850*, (Cambridge, 1989), pp. 84–189. Prices are from G. Clark, 'Labour Productivity in English Agriculture, 1300–1860', in Bruce M. S. Campbell and Mark Overton (eds), *Land, Labour, and Livestock: Historical Studies in European Agricultural Productivity*, (Manchester, 1991), pp. 215–16 and A. H. John, 'Statistical appendix', in: Mingay, *The Agrarian History of England and Wales, VI*, pp. 974–1009.

in population is to reflect a growth in agricultural output. The first is that no food was exported or imported. This is obviously incorrect, especially for the later period, so Table 1.1(a) also gives some rough estimates of net imports as a percentage of total output. This modification of population numbers by assumptions about exports and imports is essentially the procedure adopted by Deane and Cole in their estimates of agricultural output, although their population estimates have been superseded and their import figures refined.[4]

The use of demand equations overcomes the second assumption inherent in the population method which is that consumption per head remained constant over time. Crafts has pointed out that output trends based on population are inconsistent with the behaviour of agricultural prices. When agricultural prices are falling it is likely that per capita consumption will increase, and conversely, when prices are rising per capita consumption should decrease. Therefore, he estimates agricultural output by taking prices and wages into account together with assumptions about income and price elasticities of demand. His technique has been further refined by Jackson, and both sets of estimates are shown in Table 1.1(b).

Direct estimates of the volume of output have recently been attempted for some commodities by Chartres and Holderness. They are based on information recorded by contemporaries (who had no way of knowing how accurate their estimates were), assumptions about the per capita consumption of various products, and scattered information from farm-based evidence. In some cases, contemporary estimates have been revised and informed guesses have been used to interpolate the gaps. The revisions are often based on the evidence of population growth and assumptions about per capita consumption and the progress of agricultural technology which introduces a degree of circularity into their construction. Thus, these volume-output figures must be subject to quite a wide margin of error and are not independent of output estimates based on population growth. In addition, the interpolation of gaps in the time series may have the effect of smoothing over fluctuations.

2 Land productivity

National estimates of land productivity can be derived from these output figures by dividing them by the agricultural area of the country (Q/T), and the resulting estimates are shown in Table 1.2(a), using both the population method and the volume method for calculating output. The figures for the area of arable, pasture and meadow are contemporary estimates because the true acreage was not actually measured until 1866.

The only continuous series of land productivity figures available for the early modern period (Table 1.2(b)) are those referring to grain yields per sown acre which, as has already been pointed out, are far from the ideal measure of land productivity. Apart from the evidence of medieval account

Table 1.2 Land productivity estimates
(a) Land productivity based on estimates of the volume of output (£ per acre)

Year	*Population method*	*Volume method*	
	Index	*Productivity £ per acre*	*Index*
1700	100	1.91	100
1750	108	2.22	116
1800	116	2.68	140
1850	208	4.23	221

(b) Wheat yields per acre (bushels)

	S. England	*Lincolnshire*		*Norfolk & Suffolk*		*Herts*	*Hants*
		(1)	*(2)*	*(1)*	*(2)*		
c. 1300	12.0			14.9	11.5		10.8
c. 1550	13.3	9.5	8.0			9.0	
c. 1600	13.6	11.7	9.9	12.0	8.5	12.2	11.0
c. 1650	16.3	15.8	10.0	14.5	9.3	16.0	12.9
c. 1700	18.9	15.6	10.7	16.0	9.2	17.0	
c. 1750	22.0	20.0	13.5	20.0			
c. 1800	25.0	21.0	15.8	22.4		24.0	21.0
1830s		22.9	20.0	23.3	21.0	21.6	21.6
1860	28.0	31.0		31.1		28.0	27.0

Note: (1) Wheat bushels per acre (2) Wheat, rye, barley and oat yields, weighted by crop proportions and crop price relative to wheat.
Sources: (a) Output figures from Table 1.1 divided by estimates of land area from Gregory King, and from H. C. Prince, 'The changing rural landscape, 1750–1850', in: Mingay, *Agrarian History of England and Wales, VI*, pp. 30–3: (b) southern England: G. Clark, 'Yields per acre in English agriculture 1266–1860: evidence from labour inputs', *Economic History Review*, XLIV (1991), p. 457; Lincolnshire, Norfolk and Suffolk: Mark Overton, 'The determinants of crop yields in early modern England': in Bruce, M. S. Campbell and Mark Overton (eds), *Land Labour and Livestock: Historical Studies in European Agricultural Productivity* (Manchester, 1991), pp. 302–3; Hertfordshire and Hampshire: Mark Overton and Bruce Campbell, 'Productivity change in European agricultural development': in Campbell and Overton, *Land Labour and Livestock*; Paul Glennie, 'Measuring crop yields in early modern England', in: Campbell and Overton, *Land, Labour, and Livestock*, p. 273; R. J. P. Kain, *An Atlas and Index of the Tithe Files of Mid-Nineteenth-Century England and Wales* (Cambridge, 1986).

Table 1.3 Crop yields and output for 20 English counties, 1801, 1831 and 1871 (1801 = 100)

	Wheat			*Barley*		
	acres	*yield*	*output*	*acres*	*yield*	*output*
1801	100	100	100	100	100	100
1830s	150	98	149	140	94	139
1871	145	130	187	141	109	159

Note: The counties are Beds, Bucks, Cheshire, Cambs, Cornwall, Derbys, Durham, Essex, Hants, Kent, Lincoln, Northumb., Rutland, Salop, Staffs, Surrey, Sussex, East, West and North Ridings of Yorks.
Sources: M. E. Turner, 'Agricultural productivity in England in the eighteenth century: evidence from crop yields', *Economic History Review*, 35 (1982), pp. 489–510; *idem*, 'Arable in England and Wales: estimates from the 1801 crop return', *Journal of Historical Geography*, 7 (1981), pp. 291–302; Kain, *Atlas and Index of the Tithe Files*, Mark Overton, 'Agriculture', in: J. Langton, and R. Morris (eds), *An Atlas of Industrializing Britain 1780–1914* (London, 1986), pp. 34–53.

rolls (which peter out in the mid-fifteenth century) the earliest reliable direct estimates of yield were produced *c.* 1800, and thereafter there are reliable estimates for the 1830s and 1860s. Earlier figures are indirect estimates: those for Hampshire, Hertfordshire, Lincolnshire, Norfolk and Suffolk are estimated from probate inventories; while the southern England series are based on a comparison of day-work and piece-work harvest wage rates by Clark. Yields can be calculated for grains other than wheat, and so an index of the yields of wheat, rye, barley and oats, weighted by the area of each crop and its price relative to wheat, are shown for Lincolnshire, Norfolk and Suffolk. Unfortunately, sources do not permit these grain yields to be calculated in terms of the arable area, because sources do not record the area of fallow. Yield information for *c.* 1800 and *c.* 1836 is available for more counties than yields from inventories, so Table 1.3 presents yields for 20 English counties for 1801, the 1830s and 1871 (together with estimates of acreage). These are based on information from a large number of parishes although the acreage figures for 1801 are likely to be underestimates.

3 Labour productivity

Although sources are available from which 'bottom-up' estimates of labour productivity can be calculated for individual villages or farms, very few such calculations have been made. This is unfortunate because it is only by using

Table 1.4 Labour productivity estimates
(a) 'Population' method

Date	*Rural agricultural population*	*'Labour productivity'*	*Index (1700 = 100)*
1520	1.82	1.32	71
1600	2.87	1.43	77
1670	3.01	1.65	89
1700	2.78	1.86	100
1750	2.64	2.34	126
1801	3.14	2.62	141
1831	3.38	3.45	185
1851	3.84	3.66	197
1871	3.35	4.81	259

(b) Labour productivity based on estimates of the volume of output (£ per head of the rural agricultural population)

	Productivity	*Index*
1700	14.46	100
1750	19.36	134
1800	24.51	170
1850	29.77	206

Sources: (a) Population-based output figures in Table 1.1 divided by estimates of the rural agricultural population from E. A. Wrigley, 'Urban growth and agricultural change: England and the continent in the early modern period', *Journal of Interdisciplinary History*, 15 (1985), pp. 140–1: (b) Volume-based output measures divided by Wrigley's estimates of rural agricultural population.

such farm-based sources that estimates can be made of productivity measured in terms of output per worker hour as opposed to output per worker employed in agriculture. Consequently, the best estimates of labour productivity available at the moment divide the indices of output already discussed by the number of workers in agriculture. The difficulties of estimating employment levels before the first census in 1831 necessitate using a rather crude indicator of those working in agriculture which is the size of the 'rural agricultural population' as estimated by E. A. Wrigley. The results of this exercise are shown in Table 1.4.

4 Total factor productivity

In all the above calculations, the most reliable information is for population and exports. As output calculations become theoretically more sophisticated so the empirical evidence on which they rely becomes increasingly less reliable. Moreover, their sophistication comes from the application of economic assumptions about the behaviour of the economy which may not be appropriate in an eighteenth-century context. This is also the case with total factor productivity estimates for agriculture, which rely on uncertain evidence of output, price levels, and the relative shares of rent and wages in national income. They also depend on a wide range of economic assumptions about the nature of the economy, such as the existence of perfectly competitive product and factor markets, constant returns to scale, and disembodied technological change.

Crafts' estimates in Table 1.5 are based on estimates of physical output, but to overcome the difficulties of obtaining data on output, an alternative method derives total factor productivity from the ratio of input to output prices, rather than from changes in the relative quantities of factor inputs. This method assumes that any decline in output prices that was not due to a decline in input prices must be a reflection of lower costs induced by productivity change. The other estimates in Table 1.5 employ this method but Mokyr's results show how sensitive they are to the choice of price and wage series, and to the weights given to rents and wages in the calculations.

Table 1.5 Total factor productivity estimates (percentage change per annum)

	(a) Crafts ('output' method)
1761–1800	0.2
1801–1831	0.9
1831–1860	1.0

	(b) McCloskey ('price' method)
1780–1860	0.45

	(c) Huekel ('price' method)
1790–1815	0.2
1816–1846	0.3
1847–1870	0.5

(d) Mokyr ('price' method)

(i) Using Williamson's wage data and Gayer-Rostow-Schwartz prices

	(1)	*(2)*
1797–1827	0.13	0.02
1797–1835	0.37	0.27
1805–1827	−0.21	0.15
1805–1835	0.18	0.42

(ii) Using Bowley-Wood wage data and Gayer-Rostow-Schwartz prices

	(1)	*(2)*
1790–1820	0.32	−0.39
1820–1850	0.36	0.98

Note: 1 Assuming the proportion of wages at 0.75 of factor prices and rents at 0.25. 2 Assuming the proportion of wages at 0.33 of factor prices and rents at 0.67.
Sources: (a) N. F. R. Crafts, 'British economic growth, 1700–1850; some difficulties of interpretation', *Explorations in Economic History*, 24 (1987), p. 251; (b) D. N. McCloskey, 'The industrial revolution 1780–1860: a survey', in: R. C. Floud and D. N. McCloskey (eds), *The Economic History of Britain since 1700* (2 vols, Cambridge, 1981), I, p. 114; (c) G. Hueckel, 'Agriculture during industrialisation', in: Floud and McCloskey, *The Economic History of Britain since 1700*, 1, p. 192; (d) Joel Mokyr, 'Has the industrial revolution been crowded out? Some reflections on Crafts and Williamson', *Explorations in Economic History*, 24 (1987), p. 310.

Output, productivity, and changes in agricultural production

Table 1.6 draws together some of the estimates of output and productivity discussed so far, and expresses them in terms of percentage changes per annum for a variety of periods. Although there is a number of discrepancies between the alternative estimates, it is clear that, over the course of the eighteenth century, the output of English agriculture just about kept up with the rate of population growth, although growing imports of food were necessary to prevent Malthus's prophesy from coming true. From 1700 all the estimates show that output, land productivity and labour productivity were rising, marking a decisive break with the past, when sustained rises in all three did not take place.

Table 1.6 Population, output and productivity 1650–1850 (percentage change per annum)

	Population		*Land area*				*Output*			*Land Productivity*			*Labour Productivity*	
	Total	*Non-agricultural*	*Total*	*Arable*	*Sown arable*	*Meadow & pasture*	*(1)*	*(2)*	*(3)*	*(4)*	*(5)*	*(6)*	*(7)*	*(8)*
1650–1700	−0.07	–	–	–	–	–	−0.03	–	–	–	–	0.18	–	–
1700–1750	0.26	0.64	–	–	–	–	0.38	0.48	–	0.15	0.30	0.42	0.46	0.59
1700–1760	0.33	–	–	–	–	–	0.36	–	0.60	–	–	–	–	–
1750–1800	0.82	1.14	–	–	–	–	0.55	0.81	–	0.14	0.38	0.20	0.23	0.47
1760–1800	0.86	–	–	–	–	–	0.62	–	0.44	–	–	–	–	–
1800–1830	1.45	1.97	–	–	–	–	1.18	–	1.18	–	–	0.04	0.92	–
1830–1850	1.16	1.33	–	–	–	–	0.94	–	–	–	–	0.90	0.30	–
1700–1800	0.55	0.89	0.32	0.25	0.30	0.38	0.46	0.65	0.53	0.15	0.34	0.31	0.35	0.54
1800–1850	1.36	1.71	−0.09	0.57	0.78	−0.70	1.08	0.81	–	1.17	0.92	0.47	0.67	0.39

Notes:

1 Population-based method (Table 1.1)
2 Volume-based method (Table 1.1)
3 Crafts' estimates (Table 1.1)
4 Derived from population-based output method (Table 1.2)
5 Derived from volume-based output method (Table 1.2)
6 Wheat yields from counties with inventory data (Table 1.2)
7 Derived from population-based output method (Table 1.4)
8 Derived from volume-based output method (Table 1.4)

These findings prompt speculation as to the ways in which output rose, and the contribution of productivity change to those output increases.[5] The most obvious way for farmers to produce more food in the face of rising demand is by expanding the area cultivated, and it is clear that this made an important contribution to output growth in the eighteenth century. As population pressure increased after mid-century, the area of arable, pasture and meadow (T) increased by some 0.42 per cent per year. This relatively rapid rate of reclamation of marsh, heath and woodland hitherto exploited for food at a very low level of intensity, accords with independent evidence of land reclamation, particularly through the process of enclosure.

Thus, on the evidence of Table 1.6, in the eighteenth century a greater proportion of the increase in food output came from the extension of the area of arable and pasture than from increases in land productivity measured as total output divided by the area of arable, meadow and pasture (Q/T). Nevertheless, this measure of land productivity was also growing, and by the nineteenth century was responsible for all the increase in domestic output. Part of this increase was due to higher yields (QC/TC) for individual crops; but cereal output could grow even if yields and the area of arable, pasture and meadow remained constant. The cultivation of fodder crops, particularly turnips and clover, enabled the arable area to increase at the expense of pasture (Equation 7) and the area of cereals and fodder to expand at the expense of fallow (Equation 4). Aside from their direct effects on grain yields per acre, these new crops provided more fodder per acre than permanent grass or meadow, and turnips were especially important in enabling cereal to be grown on light land formerly under pasture.[6]

During the first half of the nineteenth century, the total area of meadow and pasture probably fell although the arable area rose from 11.6 to 13.7 million acres (a rate of 0.33 per cent per year), and the area under wheat grew even faster, by 0.73 per cent per annum. If the comparison between King's and Comber's figures is to be trusted, the arable area grew only slightly in the eighteenth century, although the proportion of fallow in rotations may have been falling as more fodder crops were cultivated.

The increased cultivation of fodder crops can be demonstrated from farm-based evidence, but unfortunately results for the period 1660–1850 are available only for two counties. Table 1.7 shows crop and livestock statistics from probate inventories, the tithe files, and the 1854 crop return for Norfolk and Suffolk. Although information on the proportions of fallow and permanent pasture is unavailable, the proportion of fodder in arable rotations rose at least threefold between the first 40 years of the eighteenth century and the 1830s. Although figures on absolute acreages are also unavailable, the proportion of the cereal acreage under wheat nearly doubled over the same period.

Table 1.7 Norfolk and Suffolk: crop combinations, 1660–1854

	1660–99	*1700–39*	c. *1836*	*1854*
% Grain area:				
Wheat	29.9	26.7	48.7	49.1
Rye	10.9	7.3		1.2
Barley	47.8	54.0	46.8	42.1
Oats	9.4	11.2	4.5	7.6
% Cropped area:[a]				
Legumes	16.4	11.9	27.8	25.6
Fodder[b]	13.8	15.1	51.4	45.2
Turnips	0.9	8.2	20.4	19.6
Clover	0.1	2.8	23.6[c]	19.1
Bare fallow as % cropped area			5.9	3.9
% Grass[d]				26.9
% Livestock units[e]				
Cattle	61.8	66.7		38.4
Horses	27.9	25.6		24.2
Beasts per 100 cereal acres				
Livestock	80.5	70.9		56.0
Cattle	42.7	51.1		21.0
Horses	16.6	15.5		13.6

Notes:

[a] Excludes fallow, meadow, and permanent pasture.

[b] Legumes and roots.

[c] 'Seeds'.

[d] Meadow, permanent and rough pasture (excluding clover) as a percentage of grass and arable.

[e] Livestock units (horses × 1.0) + (oxen, cows, and bulls × 1.2) + (immature cattle × 0.8) + (sheep × 0.1) + (swine × 0.1)

Source: Mark Overton, 'The determinants of crop yields in early modern England', in Campbell and Overton, *Land Labour and Livestock*, pp. 306–8.

Livestock productivity also increased. A comparison of King's guess at the number of livestock in the country *c.* 1700 with the more certain figures for the 1850s suggest that there was little growth in numbers between the two dates. Given the more certain evidence of the expansion of the arable acreage, this is corroborated by the fall in the density of livestock per cereal acre, between the early eighteenth century and the 1850s shown in Table 1.7. Yet the output of livestock products grew roughly at the same rate as the growth in output as a whole (using the volume method), so the productivity of livestock in terms of output per beast must have risen. In part this may have been due to the breed improvements of the eighteenth century which loom large in traditional accounts of the agricultural revolution, but it is also possible that new fodder crops provided better levels of nutrition. Both would also contribute to reducing the time taken for cattle to 'finish' and be ready for slaughter, thus enabling a higher turnover of animals.

These increases in output appear as an increase in land productivity measured as Q/T, but not in the more restricted index of cereal output per unit sown (QC/TC), which has usually been taken as the index of land-productivity improvement by historians. The determinants of crop productivity in this period are complicated,[7] but, from an agronomic point of view, the key to higher crop yields before about 1830 is the supply of nitrogen which was the 'limiting' factor in cereal growth.[8] Growing crops remove nitrogen from the soil and this nitrogen must be replaced if crops are to continue to be grown. The three ways of doing this are: first, by exploiting existing reserves of nitrogen; second, by recycling nitrogen removed to minimize losses from the system; and third, by finding new sources of nitrogen. The major reserve of unexploited nitrogen lay in permanent pastures, and the spread of arable husbandry at the expense of pasture discussed above could have resulted in a short-term boost in yields as the nitrogen was released. In the longer term, the crucial development for improving the efficiency of nitrogen recycling was the integration of livestock and crops. In the Middle Ages the two branches of farming were conducted separately so that there was no correlation between livestock stocking densities and cereal yields. By the mid-seventeenth century there was.[9] Convertible husbandry, whereby grass was rotated round the farm, enabled nitrogen in grass to be recycled to the arable through livestock. This trend continued into the eighteenth century but, by then, new sources of nitrogen were being added to the soil through increased acreages of legumes (mostly in the form of clover) which have the property of converting atmospheric nitrogen into mineral nitrogen in the soil.

By the 1830s the full potential of legume-based intensive husbandry seems to have been reached. Indeed, if the comparison between the yields recorded in the 1801 crop return and the evidence of the tithe commissioners is valid, then yields had risen little over the first three decades of the nineteenth century. Between the 1830s and 1850s, however, yields of wheat jumped again. It is possible that, by this time, nitrogen was no longer the 'limiting factor' in crop growth, so further applications of nitrogen may have had little effect on yields. The element most likely to have been deficient was phosphorus, and this mid-nineteenth-century rise in yields coincides with the importation of large quantities of phosphate fertilizer from overseas, dubbed by Thompson a 'second agricultural revolution'.[10] Thereafter, yields rose little until the second half of the twentieth century, when artificial fertilizers, pesticides and herbicides were to revolutionize the physical output of farming once again.

There may well have been other technological changes responsible for raising yields aside from increasing the nitrogen available to crops. Scientific plant breeding is a twentieth-century phenomenon, but farmers in the early modern period were certainly aware of the benefits of selecting seed, and it is possible that improved, higher-yielding varieties of seed were being developed. By the early nineteenth century, there is clear evidence of seed selection and the development of new strains of wheat and of barley.[11]

As yields increased by these means, the proportion of the cereal harvest remaining for human consumption grew proportionately faster (Equations 2 and 3). This is because the quantity of seed that had to be saved for the following harvest remained roughly constant, and so formed a smaller proportion of a growing cereal output. Or, to put it another way, the rate of growth of cereal yields per acre net of seed was faster than the rate of growth of gross yields per acre.

The determinants of labour productivity are equally complicated. By the mid-nineteenth century mechanical innovations were undoubtedly having some effect upon labour productivity. Cutting wheat with a scythe instead of a sickle doubles labour productivity, for example; it is further doubled with the mechanical reaper, and doubled yet again by the reaper-binder. Collins has indicated the potential labour productivity gains from these developments, and Walton has provided detailed evidence of the diffusion of agricultural machinery in Oxfordshire, but the two approaches have not been married to provide specific evidence of labour productivity changes.[12] Yet we have no evidence of new implements or machines which could have had much influence on labour productivity before the nineteenth century even though labour productivity was rising.

Increases in output per acre played a part in increasing labour productivity but they were unlikely to have been the major factor.[13] While some agricultural operations, such as ground preparation, required the same labour input irrespective of crop yields, many, such as threshing and to a slightly lesser extent harvesting, were directly proportional to yield. Thus, higher yields inevitably meant more labour was required unless harvesting or threshing technology changed. Moreover, the new fodder crops of the eighteenth century made new demands for labour.

Wrigley has suggested recently that the substitution of animal for human labour and effort was one potential source of rising labour productivity. He shows that pro rata, English farmers had two-thirds more animal power at their disposal than their French counterparts at the turn of the nineteenth-century.[14] He also suggests a number of other factors which may have affected labour productivity, including the underemployment of labour on family farms, which would depress labour productivity. This would also tie in with Allen's claim that England's superior labour productivity at the end of the eighteenth century derived from her larger farms and more rational farm layout which allowed greater efficiency in the deployment of labour.[15] Yet the issue is not clear-cut. Family farms allow fuller and more effective use of family labour, and the smaller farms of late nineteenth- and early twentieth-century Ireland (many increasingly run as family enterprises) fared better than their larger counterparts in England.[16]

Finally, labour productivity could have improved through increased labour 'efficiency'. This covers a multitude of factors but all represent some form of institutional or organizational change. The key to improved efficiency must have been through changes in employment relationships, but the effect of these on labour productivity (as opposed to other effects) has not really been explored. Changes in employment practices from the eighteenth century reduced the duration of many labour contracts from a year, to a week, or sometimes to the day. In one sense this makes more efficient use of labour because workers are not being paid to do nothing at certain times of the year, so in fact improvements in labour productivity may have been due, less to rises in output per worker-hour, than to increases in output per worker-year. Furthermore, although the introduction of turnips and clover demanded more labour, it was needed at seasons of the year when labour requirements had been low. Thus, although more labour was needed, the number of labourers required per year need not have increased. The figures for labour productivity in Table 1.4 and Table 1.6 relate output to the size of the rural agricultural population, which can be taken as a surrogate for the size of the agricultural work-force. It thus reflects output per worker per year. If workers were working more hours per year, however, then labour productivity measured as output per worker-hour would not have been increasing so rapidly. In welfare terms, therefore, the improvement may not have been so great.

It is understandable that historians have concentrated on conspicuous technological innovations, such as new crops or new machines, in their accounts of productivity change because these are clearly evident in the historical record. But the broader conception of technological change, encompassing improvements in knowledge and organization, could result in increased output for a given level of land and of labour inputs without obvious changes to farm enterprises. Such changes in farming skills and farm management undoubtedly took place but are extremely difficult to pinpoint. The supply of farming books increased from the mid-seventeenth century but, while some of these advocated best-practice techniques, others were quite bizarre in their recommendations. The provision of formal agricultural education in England did not occur until the nineteenth century, but that is not to say that levels of skill and management were not improving. By the nineteenth-century, English farmers had a growing range of literature advising them how to farm more profitably.[17]

Changes in management are often conditional on changes in the institutional environment in which farmers operate. Institutional change was thus a potentially important source of productivity growth. The eighteenth-century saw changes in farm sizes and the engrossment of holdings, changes in the conditions and security of tenures, and the replacement of common property rights by private property rights. Enclosure, and particularly enclosure by Act of Parliament, was often the means by which these changes were brought about although it was not a necessary condition for them. The impact of enclosure on productivity change continues to be debated but the most recent consensus suggests that it made only a small direct contribution to raising land productivity. On the other hand, if larger farms were more efficient in their use of labour, then it could have made a significant contribution to rises in labour productivity.[18]

Conclusion

The question as to whether these changes in output and productivity amount to a 'revolution' depends on the significance we attach to them. A strong case can certainly be made for their revolutionary significance for they represent the resolution of two dilemmas that dogged English agriculture from Middle Ages until at least the mid-seventeenth century. The first, how to raise output per acre without damaging the ecological balance, was solved by a package of biological technology which is exemplified by the Norfolk four-course rotation or one of its many variants. Turnips and clover contributed to increases in cereal yields and provided extra animal fodder, but they also enabled the arable acreage to expand at the expense of pasture, a point that has sometimes been overlooked. The second dilemma, how to raise both land and labour productivity together, was also solved by the eighteenth century, partly through these technological changes, but mainly through raising labour productivity, measured in terms of output per worker-year.

The question remains as to the timing of these output and productivity improvements. Unfortunately, the data presented in this chapter are at their least reliable when charting changes over short periods of time. From 1650 to 1700 output may well have fallen slightly and, although land and labour productivity were probably rising, their rates of growth were comparatively low. During the first half of the eighteenth century, output grew more quickly, keeping ahead of population growth, and, according to the estimates of Crafts and Jackson (but not the population- or volume-output indices), at a faster rate than in the second half of the century. Moreover, output increases owed more to productivity gains than they did to extensions to the cultivated area, whereas the reverse was true from 1750 to 1800. All the estimates of labour productivity (Table 1.6) show that labour productivity grew more quickly in the 50 years before 1750 than during the subsequent 50 years. On this evidence, therefore, there are grounds for claiming the period 1700–50 as more revolutionary than the following half-century. But both periods were outpaced by developments in the first half of the nineteenth century. Output and land productivity were growing at over 1 per cent per annum, and the growth of labour productivity approached 1 per cent during the first 30 years of the century, and, for what they are worth, the estimates of total factor productivity show faster rates of growth for the nineteenth century than they do for the eighteenth.

Notes

1 This paper was written in 1989 and some of its arguments are superseded in Bruce M. S. Campbell and Mark Overton, 'A new perspective on medieval and early modern agriculture: six centuries of Norfolk farming, *c.* 1250–*c.* 1850', *Past and Present*, No. 141 (1993) pp. 38–105; Mark Overton, *Agricultural Revolution in England: the Transformation of the Rural Economy 1500–1850* (Cambridge, 1996) and *idem*, 'Re-establishing the English Agricultural Revolution', *Agricultural History Review*, 43 (1996). For discussion of these 'agricultural revolutions' *see* J. V. Beckett, *The Agricultural Revolution* (Oxford, 1990); Mark Overton, 'Agricultural Revolution? Development of the Agrarian Economy in Early Modern England', in: A. R. H. Baker and D. Gregory (eds), *Explorations in Historical Geography: Interpretative Essays* (Cambridge, 1984), pp. 118–39; and *idem*, 'Agricultural Revolution? England, 1540–1850', in: A. Digby and C. H. Feinstein (eds), *New Directions in Economic and Social History* (London and Basingstoke, 1989), pp. 9–21.

2 J. Thirsk (ed.), *The Agrarian History of England and Wales VI, 1640–1750: Regional Farming Systems*, (Cambridge, 1984); *idem, The Agrarian History of England and Wales VII, 1640–1750: Agrarian Change*, (Cambridge, 1985); G. E. Mingay, (ed.), *The Agrarian History of England and Wales, VI, 1700–1850* (Cambridge, 1989).

3 For a general discussion of agricultural productivity in an historical context *see* Mark Overton and Bruce M. S. Campbell, 'Productivity change in European agricultural development', in: Bruce, M. S. Campbell

and Mark Overton (eds), *Land Labour and Livestock: Historical Studies in European Agricultural Productivity* (Manchester, 1991), pp. 7–17.

4 P. Deane, and W. A. Cole, *British Economic Growth, 1688–1959* (2nd edn, Cambridge, 1967), p. 65.

5 These issues are discussed in Overton and Campbell, 'Productivity change', pp. 17–28.

6 Mark Overton, 'The diffusion of agricultural innovations in early modern England: turnips and clover in Norfolk and Suffolk 1580–1740', *Transactions of the Institute of British Geographers*, new series, 10 (1985), pp. 205–21.

7 Mark Overton, 'The determinants of crop yields in early modern England', in: Campbell and Overton, *Land Labour and Livestock*, pp. 284–322.

8 R. S. Shiel, 'Improving Soil Fertility in the Pre-Fertilizer era', in: Campbell and Overton, *Land Labour and Livestock*, pp. 51–77.

9 Overton and Campbell, 'Productivity change', pp. 35, 43.

10 F. M. L. Thompson, 'The second agricultural revolution, 1815–1880', *Econ. Hist. Review*, 2nd. ser., 21 (1968), pp. 62–77.

11 Overton, 'Determinants of crop yields', p. 290.

12 E. J. T. Collins, 'The age of machinery', in: G. E. Mingay (ed.), *The Victorian Countryside*, (2 vols, London, 1981), I, pp. 200–13; J. R. Walton, 'Mechanisation in agriculture: a study of the adoption process', in: H. S. A. Fox, and R. A. Butlin (eds), *Change in the Countryside: Essays on Rural England, 1500–1900*, Institute of British Geographers Special Publication, X, (London, 1979), pp. 23–42

13 G. Clark, 'Labour Productivity in English Agriculture, 1300–1860', in: Campbell and Overton, *Land Labour and Livestock*, pp. 211–35.

14 Wrigley, E. A., 'Energy availability and agricultural productivity', in: Campbell and Overton, *Land Labour and Livestock*, pp. 323–39.

15 R. C. Allen, 'The growth of labour productivity in early modern English agriculture', *Explorations in Economic History*, 25 (1988), pp. 117–46; G. Clark, 'Labour productivity and farm size in English agriculture before mechanisation: a note', *Explorations in Economic History*, 28 (1991), pp. 248–57.

16 P. Solar and M. Goossens, 'Agricultural productivity in Belgium and Ireland in the early nineteenth-century', in: Campbell and Overton, *Land Labour and Livestock*, pp. 382–3.

17 H. S. A. Fox, 'Local farmers associations and the circulation of agricultural information in nineteenth-century England', pp. 43–63 in *idem* and Butlin, *Change in the Countryside*; N. Goddard, 'Agricultural literature and societies', in: Mingay, *The Agrarian History of England and Wales VI*, pp. 361–83.

18 M. E. Turner, 'English open fields and enclosures: retardation or productivity improvements', *J. Econ. Hist.* 46 (1986), pp. 669–92; R. C. Allen, 'The two English agricultural revolutions, 1459–1850', in: Campbell and Overton, *Land Labour and Livestock*, pp. 236–54.

2

Agriculture and Economic Growth in Britain, 1870–1914

F. M. L. Thompson

At first sight, it may seem a thankless task, perhaps even a meaningless exercise, to speak of agriculture and economic growth in the same breath when considering the case of Britain between 1870 and World War I. On the one hand, a whole school of thought among economic, cultural and imperial historians has created a general impression that there can scarcely have been any economic growth in this period, because they have been preoccupied with suggesting explanations for the onset of British economic decline.[1] On the other hand, agricultural historians have accustomed us to believe that, regardless of what may have happened to the rest of the economy, British agriculture entered into a great depression and decline which lasted, on some interpretations, until the outbreak of World War II.[2] More balanced views, however, paying proper attention to the national accounts, have shown that the British economy as a whole continued to grow in this period, and that the question marks over the performance of the economy are not those of decline, but of growth rates which were lower than those of major competitors, such as Germany and the United States, which were engaged in eroding the long lead that the British had established by the mid-nineteenth century in industry, commerce and finance.[3] In this perspective, agricultural historians need to emphasize not so much the trials and tribulations of a depressed agriculture as the ways in which agriculture adapted to an industrialized economy operating in a free-trade international economy, and to examine its relationship to the sustained growth of a mature economy.

An extreme view would be that the British economy would have been better off if there had been no agricultural sector at all. Agriculture had the lowest labour productivity and total factor productivity of any sector (except possibly domestic service), and hence its use of resources dragged down the performance indicators of the economy as a whole. Comparative advantage, it could be held, indicated that Britain should have concentrated on those manufactures and services in which it excelled, and the development of new territories and of the international economy would ensure that Britain could procure all its agricultural supplies through trading. The experiences of World War I might seem to have exposed that reasoning as more than a trifle unwise but, in the run-up to 1914, the best expert opinion held firmly that a strong navy was a sufficient safeguard against any notional vulnerability created by very large, or even total, import dependence for food supplies.[4] Some contemporaries certainly were untroubled by an economic logic that might decree the extinction of farming in Britain: Andrew Bonar Law, who dropped agricultural protection from the party programme shortly after becoming leader of the Conservative opposition, told one of his leading landowner MPs, in 1912, that he would willingly accept reliance on Canada for the whole of Britain's wheat supply 'thus rendering its cultivation in Great Britain quite unnecessary'.[5] Others professed the belief that unfettered market forces were leading to the elimination of agriculture, while deploring this result. 'Agriculture in Britain is dead', an official of the Board of Agriculture minuted in 1900, 'it only remains to give it a decent burial'; and a couple of years later, Rider Haggard, who did duty as a traditional country gentleman as well as a highly successful writer of colonial adventure stories, reported that 'owing principally to the lowness of prices . . . and the lack of labour, I take it to be proved then that in the majority of districts English agriculture is a failing industry'.[6]

If the nil-agriculture hypothesis is correct in stating that the maximum possible contribution of agriculture to economic growth is to disappear altogether, *a fortiori* a decline in agriculture relative to other sectors of the economy will make a contribution to economic growth of some size short of the theoretical maximum. Later, many were to argue that one important factor in explaining why the post-1945 growth rates of the French, Italian and even German economies outstripped the British, was that they benefited from massive transfers of labour from low-productivity agriculture into high-productivity industry and services, which the British economy was precluded from doing because these transfers had already taken place half-a-century or more earlier.[7] In the period 1870–1914, other economies received a boost from this source, notably the German, in which agriculture's contribution to GNP fell from 40 to 23 per cent, and agriculture's share of the total labour force from 46 to 37 per cent. But relatively, if not in terms of absolute quantities of output or people employed, the British economy experienced the largest boost of all, with agriculture's share of GNP and of the total labour force halved, and more than halved: the declines were from 15 to 6 per cent in the first case, and from 16 to 8 per cent in the second.[8]

So far the argument is no more than a rephrasing of the classic model of the relationship between agriculture and economic growth, in which the role of agriculture is to release underemployed resources, principally but not exclusively labour, for more productive employments. This release is possible, given that a relatively growing non-agricultural population increases its efficiency and productivity, or if alternative sources of agricultural supplies become available through imports. A third possibility, that the release of resources from agriculture can be achieved by forcing the entire population to reduce its levels of consumption, though once favoured by 'pessimists' as a description of what happened in Britain roughly between 1760 and 1830, is probably possible only in command economies. It would now be generally agreed that the first option, increasing agricultural productivity, was what underpinned the growing non-agricultural proportion of the total British population from at least the early eighteenth century to the 1840s, thus creating the necessary food supply conditions in which industrial and urban growth could take place. The growth of food imports, significant in some commodities from the late 1840s onwards, and prominent across a wide range of temperate-zone products by the last quarter of the century, created a general impression that Britain passed decisively into the second phase between 1870 and 1914, in which domestic agriculture continued to disgorge resources simply because overseas suppliers made good, and more than made good, the resulting decline in home-grown supplies.

If this was all that there was to say about the relationship between agriculture and economic growth in Britain between 1870 and 1914, then the story would simply be one of a declining agriculture unable to cope with import competition and making a passive and involuntary contribution to the growth of other sectors of the economy. Leaving aside non-competing imports of foods from Mediterranean, tropical, and subtropical climates (tea, coffee, cane sugar and many fruits being the main ones), some level of imports of competing foods (grains, meat and dairy produce) was clearly necessary because of the sheer size of the British population and the physical limits on home production under the available technology. That these physical limits were not inflexible nor unduly restrictive was to be shown, under entirely different technological and political conditions, by the capacity of British farming in the 1980s to deliver something like 75 per cent of the home demand, and produce exportable surpluses of barley and eggs.[9] When Britain moved, between 1870 and 1914, towards dependence on imports for about 60 per cent of home consumption of temperate-zone foods, clearly more was involved than just the physical and technical ceilings of British cultivable acreage.[10] When it is also observed that, in the same period, the output of agricultural products per person employed grew by at least one per cent per annum on average – a rate which compared favourably with those of other sectors of the economy – it is apparent that more was involved than a plain collapse of British agriculture in the face of import competition.[11]

The usual argument, following the opinions of many contemporaries, is that the response of British agriculture to the flood of imports was at best inadequate, and in most cases supine. The British farmer had the largest and richest mass market for foodstuffs in the world on the doorstep, and to surrender huge segments of it to farmers whose produce had to travel many hundreds or thousands of miles seemed not just careless but exceedingly and almost inexplicably feeble.[12] To surrender to American, Canadian, Australian, Argentinian, Russian and Indian wheat was one thing, although even there some explanation beyond that of radically lower costs of production, which the British could never match, is called for, because British wheatlands were two or three times more fertile than any of the others. To surrender to Danish bacon and eggs or Dutch butter and cheese was another matter, given the broad similarities of soils, climate and farming traditions; failure to match them in quality and price was taken to indicate either entrepreneurial failure by British farmers, who did not see the market openings or did not know how to achieve the uniformity of product quality necessary for their exploitation, or the rigidities and obstacles imposed by the British agrarian structure, which forced the farm sector to support an expensive and idle class of landowners.[13]

There are many deeply emotional and irrational features in this picture, overlying such lesser truths as that British bacon failed to match the consistently high standards of the Danish rasher and that British farmers were slow to grasp that eggs and poultry were something more serious than a hobby for their spouses, shortcomings which did, indeed, suggest lack of entrepreneurial vigour and imagination. At the general level, however, there was nothing reprehensible or economically irrational about the British eating foreign bread and butter, any more than there was anything reprehensible about foreigners wearing clothes made of British cottons or woollens or riding in trains pulled by British locomotives. As for the landlord class, economic rent, in the Ricardian sense of the surplus production of the more fertile and better-located land over the marginal land in which cultivation is ensured by existing levels of effective demand, exists under any conditions of tenure and is present where agriculture is in the hands of owner-farmers just as it is in an explicit structure of landlords and tenant farmers. It may well have been that, in the international economy of the late nineteenth century for free-trade Britain, the marginal 'no-rent' agricultural land, that theoretically set the levels of rent yielded by the more favourably endowed land, lay on the agricultural frontiers of the new worlds and not within

Britain itself. It is probably true that the claims of many agricultural apologists, that 'pure' economic rent had vanished in Britain by the end of the century, leaving the rent paid by tenants as simply a payment for the capital invested by landowners in creating the farmland and its physical equipment, were greatly inflated.[14] If some element of 'pure' rent was still contained in the rents actually paid by farmers to landowners in this period, then the serious question is not whether the landowner class was a redundant millstone around the neck of British agriculture, but whether imperfections in the market mechanism – customary, institutional, social or political – prevented the adjustment of that element far enough to preserve, or create, 'a level playing field' on which British farmers could compete with others. Likewise with import penetration of the British food market, the serious question is not whether it was a sign of weakness or failure to have allowed any penetration at all, but whether the actual level of penetration in particular products was larger (or smaller) than was consistent with the optimum welfare of the British people as a whole.

It is easier to pose such questions than to answer them empirically. Given that the prices of the majority of foodstuffs in British shops were set by international markets and were not influenced by the varying, often small, proportion of home-grown supply, it is likely that the welfare of the British people, as consumers, was being maximized. Whether the welfare of the small and declining proportion of the British people who were food producers could have been improved by producing more, at the prevailing prices which they could not influence, depends on the scope for reducing costs of production below the levels actually achieved, and raising the quality of entrepreneurship and farming techniques. The principal cost was labour. Farmers may have dearly wished to reduce the price of labour, and, during the 1880s, did succeed in imposing some short-term reductions in wage rates. But increasingly the wage rates of agricultural workers were externally determined by the non-farm influences of alternative industrial and urban employment, or emigration, a situation earlier limited to agricultural workers within the northern and midland industrial regions. The result was that, over the period as a whole, money wages rose by at least 15 per cent and real wages by well over one-third.[15] The only effective way in which farmers could reduce their labour costs, therefore, was by reducing the number of labourers they employed. This they proceeded to do, complaining the while that labour shortages were being inflicted on them by a drift from the land caused by the lure of bright city lights. The number of hired workers fell by one-third between 1871 and 1911, a larger decline than in the agricultural work-force as a whole since the numbers of both farmers and farmers' relatives working in agriculture remained fairly steady. Very roughly these figures suggest that farmers reduced their total money wage bill by about a quarter. In part this was achieved by simple abandonment of labour-consuming tasks, giving up some of the fringe activities in hedging and ditching, and converting arable to pasture or leaving land to tumble down, actions which contributed to the much-noticed neglected, unkempt, and derelict appearance of stretches of the countryside, but which were nonetheless rational responses to market forces. In part, however, the reduction in labour was achieved by the adoption of labour-saving machinery, the substitution of horses for men and women which was concomitant on mechanization, and, less easy to substantiate, the substitution of fertilizers for labour.

In the 1850s and 1860s the British may have been slow, in comparison with the Americans, in adopting farm machinery, particularly mechanized harvesting – probably explicable, despite ingenious arguments concerning the obstacles of threshold farm and field sizes, on straightforward grounds of abundant supplies of cheap labour, especially of harvest labour. Later in the century, however, particularly in the 1890s, adoption of machinery, including the recently developed (American) combined reaper-binder, went ahead rather rapidly, and by the eve of World War I, British agriculture was easily the most highly mechanized in Europe.[16] Curiously, since grain growing was the sector of British farming under most pressure and its contraction was the most highly publicized feature of agricultural decline, the most eye-catching parts of this mechanization were concentrated on the cereal sector: in threshing where machines were already normal before 1870, but were subsequently widely replaced by larger and more sophisticated models, and in harvesting and stacking, where the machinery was generally of post-1870 development. Thus, the shrunken British cereal sector, that survived the onslaught of imported grains, survived because, on suitable soils and in suitable locations, British cereal farmers proved themselves capable of living with the new regime of low prices by cutting their production costs.[17] Some of the new machinery was applied to grassland management, especially to mowing and to haymaking, but very little, aside from barn machinery used in fodder preparation such as chaff-cutting or turnip-slicing, related more directly to the livestock side of farming. Barn machinery was largely steam driven, although some farms used water power and perhaps rather more used horse-gins; almost all the field machinery was horse-drawn. The impact on the horse population is only palely reflected in the overall movement in the number of horses employed in agriculture, which increased by 15 per cent between 1871 and 1911. In the same period, however, the total arable acreage declined by one-fifth, and if it is assumed that all the farm horses were employed on arable operations, then the intensity of their use increased sharply, by 45 per cent, from 5.1 horses per 100 arable acres to 7.4; a similar, slightly higher, rate of increase is yielded from the probably more realistic assumption that at least two-thirds of the horses were employed on the arable, which gives lower horsing ratios of 3.3 per 100 arable acres in 1871 and 4.9 in 1911. Whichever way one looks at it, the decline of one-third in the hired labour-force was offset by a rise of approaching one-half in the effective horse-force, a rise which, incidentally, implied an improvement in the average skill level of the labour force, because horsemen were among the most skilled agricultural workers.[18]

While farmyard manure continued (along with town stable manure) to be the principal source of fertilizer for farming and market gardening, farmers had begun to supplement it with purchases of 'artificial' and manufactured fertilizers from the 1840s, and, by the early 1870s, were spending £6 million a year on an array of these substances. To farmers, their purchases of fertilizers appeared as an easily visible part of their costs of production, while the costs of farmyard manure were likely to be concealed or ignored. From this perspective farmers made sharp reductions in their costs after the early 1870s, spending about £3.5 million a year on fertilizers in the early 1890s and topping the £5 million a year mark only in the last couple of years before 1914. This, however, was largely an effect of great reductions in the prices of fertilizers, and a different picture is presented by the physical quantities purchased and used. Initially, farmers economized on their purchases, and in the ten years following 1881, the total tonnages used were about 25 per cent down on the pre-1877 levels: this decline was not dramatically greater than the decline in the total acreage under tillage, which was the farmland to which fertilizers were applied, but even so it did signal a clear retreat from pre-1877 application rates and, as such, was one of the signs of the abandonment of high farming practices. From the early 1890s, consumption began to increase again, and by 1896 the total tonnage used surpassed the pre-1877 peak; thereafter the tonnages used grew steadily year by year until a surge carried them to nearly double the pre-1877 peak in 1912–13.[19]

Even the 1.5 million tons of fertilizers purchased by British farmers in 1912–13 appears insignificant when set alongside the total area of land being farmed – 32 million acres of cultivated land (crops and grass, excluding rough grazing), or 14.5 million acres of arable land. Nevertheless, these fertilizers made an important contribution to agricultural production, and consideration of particular varieties and their uses suggests why. In the early 1870s, the fertilizers in common use were guano (supplying phosphorus, nitrogen and some potassium, all three of the most useful chemical aids to fertility), Chilean nitrate (nitrogen), and superphosphate (phosphorus). By the eve of World War I, guano had almost dropped out, imports having declined to a trickle, while sulphate of ammonia (nitrogen) and basic slag (phosphorus and lime), had joined the ranks of fertilizers used in bulk: there was also a regular import of kainite (potassium) from Germany, which had a world monopoly of the supply, but it was too small to make a significant addition to the supply of potash naturally available from farmyard manure and ashes. Superphosphate was already well established in the third quarter of the century as the most popular of these fertilizers, used in the largest quantities, and it retained this position even though farmers came to spend equivalent sums on nitrates plus sulphate of ammonia by 1914. The tonnages of superphosphate used in the early 1870s were sufficient to provide an average application of about 30 pounds per acre of cultivated land (crops plus grass), a mere sprinkling if measured against contemporary recommendations to use 1½ to 3 hundredweights an acre for cereal crops and 4 to 5 hundredweights for roots. By 1912–13, the increased tonnage being used gave an average application rate that had been nearly doubled, to about 55 pounds per acre. In addition, in the early 1870s, the equivalent of 8 pounds an acre in phosphates had been supplied by the guano being used; while the guano had all but disappeared before 1914, as a source of phosphate it had been more than replaced by basic slag, which, after a slow start, British farmers were using at the rate of 17 pounds an acre. This suggests that overall phosphate inputs had grown from 38 pounds an acre to 72 pounds, an average use which still remained far below the guidelines of best farming practice. Nevertheless, not only was the rate of increase, mainly after 1896, very impressive, but also it had raised average utilization into the same league as the British farming practice of the 1980s when, on average, the equivalent of about 80 pounds of superphosphate per acre were used.[20]

If the phosphate performance of British farmers was good, their nitrogen record was uninspiring. Although the sources of purchased nitrogen inputs changed over the period, the amounts used per acre, averaged over the total cultivated area, remained almost constant. In the early 1870s, the equivalent of some 7 pounds an acre came from the guano being used and another 7 pounds an acre from the Chilean nitrate: in 1912–13 just over 7 pounds an acre came from sulphate of ammonia and about 9 pounds an acre from the imported nitrate. The contrast with the farming practice of the 1980s, in which the British used 76 pounds an acre of 'pure' usable nitrogen, the equivalent of about 380 pounds an acre of imported or manufactured nitrates, highlights the general difference between pre-1914 agricultural techniques and the nitrogen-intensity of modern farming, rather than any backwardness of pre-1914 British farming in particular.[21] There were some good reasons, and some excuses, for the reluctance of British farmers in the pre-war period to use more nitrogen. It can be argued that the levels of nitrogen in the soils used for the main nitrogen-using crops, cereals and roots, were already very generally as high as could profitably be used by the available varieties of those crops, and were supplied organically by growing nitrogen-fixing plants in the normal rotations – clover, lucerne, peas and beans – and by farmyard manure. Certainly, the available strains of wheat, barley, oats and turnips were prone to nitrogen-sickness (mangolds were less so), and excessive nitrogen caused excessive straw growth and bolting, lodging and poor harvest conditions. On these grounds, it can be maintained that the role of purchased nitrates was simply to top up local deficiencies in the organic supply, and perhaps to stimulate extra growth of ryegrass and meadow hay which responded well to the chemicals. On this hypothesis there is no evidence of lack of enterprise by British farmers on the nitrogen front; the failure, if any, was a scientific one, in that plant breeders did not develop nitrogen-friendly strains of the major grains, but that failure was global and not specifically British. On the other hand, pre-1914 nitrates were expensive, at £10 to £13 a ton, while basic slag and superphosphate were cheap, at £1.50 to £2.50 a ton, and cost-conscious farmers, with limited ideas about their total outlays on fertilizers, might well have been more chary about buying large quantities of the more expensive article regardless of the dictates of agricultural chemistry. Those contemporaries who criticized British farmers for their slowness in taking to sulphate of ammonia in the 1890s, in contrast to the Germans, must have thought that there was scope for the profitable use of more nitrogen in British farming, and ascribed the situation to the farmers' conservatism and lack of initiative. The fact that the Germans applied most of the sulphate of ammonia to sugar beet, a crop that they could grow only because it was protected and subsidized and which the British could not grow in free competition with cane sugar, provided an excuse for the unenterprising British behaviour, but not, it was thought, a complete explanation.[22]

The nitrogen record of British farmers thus admits of two contrasting views: either it showed a conservative, prejudiced, unenterprising and penny-pinching reluctance to venture into the promising realms of large applications of artificials; or it recorded sensible and rational behaviour, a shade tight-fisted perhaps, but fundamentally extremely prudent in the context of free trade and the limitations of botanical knowledge. The balance of argument favours the second view. By contrast, in their use of the third main plant nutrient, potash, British farmers scarcely had a record at all apart from the supply organically provided in farmyard manure and ashes. This may well be put down to the fact that the only commercial source of mineral potash of agricultural quality was the Stassfurt mines in Saxony, and that the bulk of their output was used at home in German farming. Britain did begin to develop an import of this mineral, kainite, which contained 12 per cent of potash, from the late 1880s: estimates were of 30,000 tons in 1892, 85,000 tons in 1905 and perhaps 180,000 tons in 1913, most of which was stated to be sold direct to farmers.[23] The peak 1913 import would have provided no more than 1.5 pounds of potash per acre of agricultural land, and, even if the organic sources supplied ten times as much, which seems improbable, farmland as a whole would have remained distinctly short of potash.

In practice, however, these fertilizers were not spread evenly and indifferently over all farmland, but tended to be targeted at specific crops. Nitrogen, for example, was required by all crops except the nitrogen-fixers, but was focused largely on grain crops. Potash was used particularly on potatoes, and to a lesser extent on mangolds, and the fact that Germany grew about six times as many potatoes as Britain went a long way to explain that country's much greater use of potash.[24] Indeed, if no more than half the kainite tonnage imported into Britain was applied to the potato crop, the potato land would have been receiving a very respectable dose of 50 pounds of potash per acre. Similarly, although most crops could benefit from phosphate, in practice, about half of farmers' purchases of superphosphate was applied to the root crop. On that basis, given that the total acreage under roots declined by 10 per cent between 1872–6 and 1892–6, and by 30 per cent over the whole period 1870–1914, farmers increased their use of superphosphate per acre by one-third in the first subperiod, and threefold over the whole period.[25]

The big trouble with all these increases in the use of fertilizers between 1870 and 1914 is that they had no apparent effect on the yields of the chief crops: although subject to seasonal fluctuations, sometimes considerable, due to weather conditions, the overall trend in the yields per acre of wheat, barley, oats, hay, roots and potatoes was completely flat.[26] Yet farmers were not silly, and they would not have gone on spending large amounts of money on fertilizers – more than £5 million annually – if these had no effect on output. The explanation is that increasing applications of fertilizers were needed simply to sustain existing yield levels. The reasons for that, in turn, were probably mainly to do with economizing on labour. Departures from traditional four- or five-course rotations in the interests of producing a greater proportion of cash crops and of saving labour, meant that some of the fertility previously supplied by green crops and folding sheep on the roots was lost and needed to be replaced by more fertilizers.[27] Reducing labour inputs on tillage crops would tend to reduce their yields by encouraging greater weed growth: nowhere would this have been more apparent than with root crops, and it seems likely that the yield-reducing effects of economies in the highly labour-intensive processes of hand-hoeing and singling were counteracted by the yield-increasing effects of the dramatic increase in the application rate of superphosphates. Running hard to stand still is not a very satisfying occupation, but that is what farmers seem to have achieved through their use of fertilizers. It is, at any rate, a form of exercise, and requires more energy and initiative than sitting passively on one's hands and waiting to be overwhelmed by adversity.

Although an essential attribute of fertilizers is to economize on land by enabling a greater volume to be grown on a given area, their primary economic function is to save labour by enabling a greater volume to be grown with a given amount of labour or, as in the pre-1914 period, the same volume with a smaller amount of labour. Thus, farmers cut their labour costs not only by letting land go out of cultivation or by converting arable into pasture, which could be interpreted as passive responses to import competition, but also by mechanization, increased use of horses, and fertilizers, which were active responses and which succeeded in maintaining the output per acre of the reduced cropped acreages, while at the same time rewarding the labour-force with rising real wages.

From the farmers' point of view rent, after labour, was the largest item in their costs of production. It is well known that rents fell sharply, by about 20 per cent, in the 20 years following the early 1870s, and by around one-quarter taking the period 1870–1914 as a whole. Whether they were reduced fast enough and far enough, through some combination of remissions, abatements, uncollected arrears and outright reductions, in years of acute crisis and very sudden price falls, is more than doubtful. Certainly, in the crisis years 1879–80, rents actually paid did not decline as much as the value of output, and farmers' incomes rather than either rent or wages bore the brunt of the recession. Right through the most difficult years until the early 1890s, it is possible that the main squeeze was on farmers' incomes: some estimates suggest that rent remained a constant proportion of agricultural output while the share of wages increased; others suggest that both rent and wages comprised an increasing proportion of total factor income, and that farmers' incomes were sharply reduced, to little more than half their earlier level in money terms. After the early 1890s, these movements went into reverse, rents and the total wage bill declined as proportions of total output and of total factor incomes, and farmers' incomes increased: so much so that, by 1909–13, farmers' incomes not only took a larger share than ever before, but were larger in money terms than they had been at the start of the period so that, given the considerable fall in the general price level, farmers' real incomes may have been about 25–30 per cent higher in 1909–13 than they had been in the early 1870s, a gain comparable to that of their labourers although enjoyed in a different subperiod.[28]

Unlike economists, farmers did not think in terms of net farm output (= total factor income) and its division between the three factors of production, but in terms of the total costs which they incurred and which had to be covered by the sales of their produce before anything was left over for their own income. From this perspective, rent declined from 31 per cent to 24 per cent of their total costs, and purchased inputs – of feeds, fertilizers, machinery and agricultural services – moved up to become the major item, increasing from 35 to 42 per cent of total outlays.[29] If the landlord class and the rents which sustained it were a burden imposed on British agriculture by force of law, tradition, political power, and social convention, and an 'artificial' handicap that prevented British farming from performing as well as others, then all these measures of the movement of rents converge in showing that they were a shrinking millstone.[30] The absolute and relative decline of rents at the level of the national accounts was, however, of much less significance than the enormous regional variations in the movement of rents. Thus, while rents in England as a whole declined by 17 per cent between 1872–3 and 1892–3, actual *increases* were registered in Westmorland, Cornwall, Cheshire, and Cumberland; and over the whole period 1872–3 to 1910–11 rent movements ranged all the way from a fractional gain in Cheshire to declines of nearly one-half in Essex and Suffolk, set against an overall national decline of 27 per cent.[31] Far from moving in a uniform fashion, rents behaved selectively and sensitively. In terms of the market-place they were sensitive to the differing profit-and-loss experiences of different kinds of farming. Broadly speaking, rents declined most steeply in the grain-growing counties which were most exposed to import competition, and declined least – amounting to increases in real terms – in those counties in which the farming types were least challenged by imports. That landlords simply charged what the market would bear, evidenced in the different levels of rent farmers of different types were willing and able to pay, is perhaps no surprise. What it meant was that rent was performing its economic function with reasonable discrimination and efficiency, and was not behaving as some kind of arbitrary tax on farming enterprise.

Most of the active steps that farmers took to reduce their costs of production, in their use of labour, horses, machines and fertilizers, and also, indeed, it could be argued the differential rent reductions of which they were more passive recipients, were focused on tillage operations; only a small part of the machinery, and after *c.* 1900 the fertilizers, were dedicated to grass. It is true that a proportion of the tillage operations was aimed at producing fodder for livestock: the entire root acreage, the bulk of the oats crop (mainly for horsefeed), and fractions of the wheat and barley, the use of which as feed grains increased substantially over this period. On an acreage basis, one-third of the total tillage acreage was devoted to fodder crops in 1873, rising to just over one-half in 1911; if the acreages destined for horsefeed are deducted, on the grounds that horses' oats were turned into power and not into saleable produce, these fodder-crop proportions are reduced to less than one-quarter and one-third in 1873 and 1911.[32] A very high proportion of these cost-reducing efforts, therefore, had the production of grains for the market – chiefly wheat and barley – as their purpose. Such a conclusion may seem perverse in the light of the conventional wisdom that it was above all cereal farming that was battered and prostrated in the Great Depression, but it is not inconsistent with the evidence that much wheatland was devastated and many cereal farmers ruined by the suddenness and depth of the price fall and the size of the surge in imports, especially in the crisis years 1879–80. Yet, after the deluge, wheat and barley cultivation did survive in Britain, and farm horses continued to live on home-grown oats in spite of the availability of a cheap imported substitute in maize. This survival took place on reduced acreages cultivated at lower costs. It is hard to see why this form of response to 'prairie competition', under a free-trade regime, should be considered less vigorous or efficient than the response of French or German farmers sheltering behind tariffs which, in theory, are normally held to protect inefficiency.

The conventional wisdom is that British agriculture did not so much confront import competition as run away from it and take refuge in products which remained naturally protected: milk, fresh meat and potatoes were the chief foods in which the home market was preserved for the British farmer by distance, perishability, or low value/bulk ratios. In 1909–13 imports accounted for zero, 30 per cent, and 6 per cent of these three articles.[33] The scale of the switch into the products enjoying natural protection from import competition is demonstrated by the decline in the contribution of the three cereals to gross agricultural output from more than one-third in the early 1870s to barely 10 per cent in 1909–13, and the increase in the contributions of meat (beef, mutton and pork) from 30 per cent to 38 per cent, and milk (liquid plus dairy products) from 12 to 17 per cent. The change is highlighted by the way in which wheat and liquid milk practically changed places as leading products of British farming: at the start of the period, income from wheat was five times that from liquid milk; at the close, liquid milk was three times as large as wheat.[34]

The move into milk production took place in new areas that were not traditional dairying counties, such as Berkshire, Essex, Hampshire, Hertfordshire and Kent, as well as in the traditional dairy leaders like Cheshire, Lancashire, Derby, Staffordshire and Somerset. It was plainly one of the major adjustments made by British farmers. Yet the concentration on selling fresh milk has excited some criticism as a tame surrender of the valuable home market for butter and cheese to imports.[35] It is true that, by the early 1900s, little butter or cheese, apart from specialities, were made in Britain except in places that were remote from speedy and reliable transport to urban markets; the proportion of total milk production devoted to butter- and cheese-making had fallen from more than two-thirds in the early 1870s to under one-quarter, and imports supplied three-quarters of the cheese and 87 per cent of the butter consumed.[36] It is also true that Britain lagged well behind many other countries – Denmark, the United States, New Zealand,

and even Ireland, for example – in the adoption of what new and cost-effective technology was available in the dairying line: principally mechanical separators and a range of quality-control devices in cheese factories. The point was, however, that even Ireland, although close to the British consumer market, was not close enough to be able to supply it with fresh milk, and therefore had to keep abreast of the most efficient techniques for converting milk into less perishable forms. British farmers were under no such pressure, and could well afford to neglect innovations in butter- and cheese-making. They did not exactly sit still and pocket the rewards of a rough doubling of the per capita consumption of liquid milk multiplied by a doubling of the number of people drinking it, a fourfold increase in the size of the milk market which was virtually a present to the farmers from demography and rising real incomes.[37] But the increase in the size of the milk market was so great that it more than absorbed all the extra milk that British farmers could produce, and, because the farmer normally received 20 to 40 per cent more per gallon for liquid milk than for its equivalent in butter or cheese, there was little obvious incentive to produce even more milk for the sake of making it into butter or cheese. Technically, it would have been possible to increase total milk production over the period 1870–1914 by more than the actual 75 per cent increase, with the aim of turning the extra gallonage into butter or cheese and making a serious effort to compete with the Danes and others, but it would have been extremely hard to convince the farmers on the economics of such an exercise.[38]

The milk story did include some supply-side changes, the most important being developments in the efficiency and availability of rail transport. On the farm, farmers entering the liquid trade presumably adopted the simple, but reasonably effective, cold-water coolers that were available by the 1880s, because it was important to cool milk rapidly before sending it on a journey of perhaps hundreds of miles or it would lose its freshness; but that was virtually the extent of dairy innovation, because reliable and economic machine-milking was not developed until the inter-war years.[39] The major part of the increase in total milk output was achieved simply by increasing the number of dairy cattle, but there was also an increase of some 28 per cent in the average annual milk yield per cow. Half that productivity gain, it has been estimated, was attributable merely to the structural changes from butter-and-cheese to liquid milk which permitted a lengthening of the milking days per cow per year; the other half was more directly the result of the effort and initiative of farmers. This half was divided in unknown proportions between improvements in breeding and improvements in feeding, and, because there does not seem to have been much progress before 1914 in moving away from the dual-purpose (milk and beef) Shorthorn as the mainstay of the dairy herd, the primary emphasis was probably on the feeding – meaning more cake and, possibly, better grass.[40] This seems to have been the measure of the farmers' own contribution to raising yields and output in the dairy industry.

If the milk trade was not literally handed to the farmers on a plate, it did not require a great deal of enterprise on their part to collect it. Much the same can be said of the meat trade. British livestock farmers more-or-less grew accustomed to the arrival in their home market in the 1860s and early 1870s of live imports of meat on the hoof from Continental Europe and the United States, which reduced their market share without denting their total sales; but they were, initially, rattled by the threat of chilled beef from the United States, which appeared in the late 1870s, and frozen beef and lamb from Australasia and Argentina from the 1880s. It did not take them long, however, to learn to live with these imports too, as it became apparent that fresh meat, and the chilled and frozen varieties served different segments of the consumer market. The result was that, while the total consumption of meat more than doubled down to 1914 (and, as with milk, both consumption per head and the number of consumers increased), and British farmers steadily lost market share, in the fresh-meat sector taken by itself, they did much better. They lost ground steadily until 1890 as imports of live cattle from the United States and Canada surged upwards, but still increased their own total production of meat, particularly of beef. After 1890 these imports ceased to grow and remained largely on a plateau while supplies of home-grown beef continued to rise. From 1909, the trade in North American cattle started to decline, because the North American home market increasingly consumed all the supplies until, by 1913, it was virtually extinct, and the vast landing wharves, lairages, slaughterhouses and stores for foreign cattle (which, under health and veterinary regulations, had to be killed at the port of entry) at Birkenhead, Liverpool, London, Glasgow and elsewhere, lay empty. The fresh-meat field was ultimately left to British farmers, not because they conquered it but because the Americans withdrew to more profitable domestic markets.[41]

The meat story was perhaps less inglorious and unimpressive than this bald account suggests. The slight overall increase of about 10 per cent in the total production of meat of all kinds (in tonnage) between 1870 and 1914 concealed a decline in the output of mutton and lamb, almost entirely the result of the steep fall in the number of 'arable sheep' which was a necessary side-effect of the decline in arable acreage and its associated corn- and-sheep farming systems, and an increase of one-fifth or more in the output of beef; the output of pork remained stable, fluctuating around its own three-year pig cycle.[42] The increase in the output of beef, however, seems to have come purely from increasing the number of cattle, including the dairy cows which were sold off as low-grade cow beef when past milking; there is no evidence of any general increase in the carcass weight of beef cattle after the 1850s.[43] British farmers, in other words, seem to have bumped up against some kind of beef productivity ceiling as early as the 1850s – much as they reached a technical ceiling in the yields per acre of grains and roots – and could look for efficiency gains only through economies on the labour used in stock management or from savings on the feed bills.[44]

There was scope for cost savings on both counts, although the lack of detailed data on the amount of labour used in looking after stock, especially cattle, makes it difficult to do more than conjecture how it was done. At first sight it might seem paradoxical that, between 1870 and 1914, the total feed bills for British livestock increased by a good deal more than the total head of animals. This was the effect of the farmers' responses to the fall in the prices of purchased feeding stuffs, chiefly of feed grains and oilcakes, both mainly imported materials. Their prices fell by one-quarter to one-third, and farmers responded by buying more and more, practically doubling the actual tonnages involved. The total head of livestock meanwhile increased by only 12 per cent, the net effect of a 24 per cent increase in the number of cattle balanced by a 12 per cent decline in the number of sheep.[45] By 1914, feeds had grown to be the third biggest item in costs of production, after wages and rent, and amounted to about one-fifth of the whole. Much of this feed was a by-product of processing industries – wheat offals from flour milling and brewers' grains left over from making beer – and the quantities consumed by farm animals were a reflex of the growth in human consumption. The most costly of the feeds, however, were the oilcakes specially manufactured for the purpose, a joint product of the oil-pressing industry, and their consumption more than doubled in the period. Although the price of oilcakes fell, they remained, at around £7 a ton just before 1914, an expensive feed which farmers reserved largely for their work-horses, stall-fed cattle and milking cows. On the assumption that the entire supply of oilcakes was fed to these animals, their annual consumption per head increased by 60 per cent, and, in spite of the price fall, this raised the annual cost per head by about 20 per cent.[46]

It is inherently improbable that farmers would give their animals more to eat, at a higher cost per head, unless they were rewarded by obtaining greater produce. With milking cows, increased rations did, indeed, contribute to an increase in milk yields. But because there was no increase in the average weight of beef cattle, the object of giving them increased rations of concentrates must have been either to induce earlier maturity, or to provide a substitute for more expensive feeds grown on the farm. The explanation of the increase in the use of purchased feeds, especially oilcakes, was quite likely to have been a mixture of all three factors: more milk per cow, earlier maturity of beef, and substitution for bulk feeds, particularly roots. All three were labour saving, either in the labour that did not have to be employed in cultivating the roots and other fodder crops that were abandoned, or in the larger produce of milk or beef per worker employed in looking after the stock. In addition, in dairying the switch from butter- and cheese-making into the liquid milk business was a move towards a less labour-intensive operation on the farm (although no doubt more labour was needed off the farm to handle milk distribution).

These developments showed up in the record of labour productivity. The measurement of labour productivity is not a simple, or precise, matter. In agriculture, the divisor may include three distinct elements: hired workers, farmers' children and other relatives who were not paid conventional wages, and the farmers themselves; while, before 1921, no information was collected that would enable the contribution of seasonal and part-time labour to be quantified.[47] In theory, the dividend ought to be the net farm output, that is, the gross output sold off the farm minus the non-farm inputs. Although it is possible to estimate the value of non-farm inputs for the United Kingdom as a whole, and with rather less confidence for the national subdivisions of the UK separately, there is no way of disaggregating national totals to county level. Disaggregation is essential to capture regional variations in labour productivity, but it can be performed only with gross farm output, which therefore has to do duty as the sole available, if imperfect, dividend in estimating changes in productivity in different counties or regions.[48] Using the widest definition of the labour-force, to include all the farmers, and estimates of gross and net output derived from Fletcher for 1867–71 and from Dewey for 1909–13, gross output per head, in Britain, increased by roughly one-third, and net output by a little over one-fifth. Clearly, not all the farmers were, in fact, working farmers; indeed, they were probably a minority so that output per head of the effective work-force was likely to have increased by more than this.[49] A narrow definition of the labour-force as consisting of the hired workers alone, a patent underestimate, produces an increase of gross output per head in England and Wales of just under one-half.[50] The difference between the results of these two measuring rods suggests very strongly that working farmers and their families comprised a growing proportion of the total effective work-force, and that one way in which farmers handled the reduction in their hired labour was to work harder themselves.

Independent estimates have been made of gross farm output, county by county, to measure regional variations, because the Census occupation tables also record the labour-force at the county level. These output estimates are not comparable with the Fletcher and Dewey estimates of national output because, when aggregated, they suggest that national output, in current prices, was 25 per cent lower in 1911 than it had been in 1873, whereas the

Fletcher and Dewey figures suggest an 8 per cent increase; but they provide a basis for gauging the relative positions of individual counties in terms of output and productivity.[51] What they show is that output per head of the hired labour-force, measured in current prices, exhibited an enormous range of movements between 1873 and 1911, varying from increases of 50 per cent or more in Somerset, Cornwall and Lancashire, to falls of 20 to 30 per cent in Lincolnshire, Essex, the East Riding, Suffolk, Huntingdonshire and Norfolk. The range is somewhat narrower, with the groups of counties showing major increases and major declines remaining the same, if wider definitions of the labour-force are used.[52] The clear implication is that labour productivity increased dramatically in those regions in which farming already had a strong emphasis on dairying and meat at the beginning of the 1870s, and fell most steeply in the traditional cereal-growing regions. The divergent trends are in large part, however, a simple reflection of divergent movements in livestock product and cereal prices over the period, coupled with the fact that grassland farming with grass-fed stock was inherently less labour intensive than arable farming. In 1873 the actual value of output per head in Lincolnshire, the East Riding, Huntingdonshire, Norfolk and Suffolk was very similar to that in Cheshire, Lancashire, Shropshire, Somerset and Cornwall; by 1911 it had fallen to little more than half their levels.

This does not mean to say that farmers and labourers in the eastern and arable counties were idle or unproductive in comparison with their western counterparts, or that their farming operations were misdirected at cheaper or less profitable products. The physical volume of output, and hence the real fruits of labour, can to some extent be captured by measuring output in constant prices. Using 1911 prices, this exercise shows that real output per head increased throughout England between 1873 and 1911, by between 10 and 25 per cent in the group of eastern counties, and between 66 and 100 per cent in the western group, thus leaving the relatively superior performance of the west untouched but acknowledging an increase in the volume of output per head in the east.[53] The volume of output measured in this way is simply the result of pricing the different items which went to make up the basket of farm produce in the different counties or regions, and hence does not reflect the entrepreneurial effort which went into altering the contents of the basket. In effect, the farmers of the western side of the country, when faced with the great price fall and the challenge of import competition – from which they were, in any case, largely sheltered – only had to do more of what they were doing before. Their production of cereals was already small in the early 1870s and it was allowed to dwindle away to insignificance; their production of meat and dairy produce was already large – around two-thirds of total output by value in Cheshire, Lancashire, Somerset or Cornwall, for example – and it was increased, to about three-quarters of total output. Managerial and technical changes were involved, in switching from butter or cheese to liquid milk, and in feeding practices: but no decisive change in direction. By contrast, the farmers on the eastern side typically depended on wheat and barley for more than half their total output in the early 1870s, and, although they did run livestock as an integral part of their farming systems, livestock products formed about one-third of output. Their response to the challenge of the times was to change their systems. Cereals were, indeed, retained, run down to about one-third of output, but they were to a considerable extent uncoupled from traditional four-course, sheep-and-barley regimes, and the livestock side of farm enterprises was expanded as one regulated by external supplies, not by the volume of farm-grown fodder. The result was that livestock products grew to one-half or more than a half of their total output. The contrast in the scale of change and adjustment is highlighted by the fact that the importance of milk, in relation to total production, increased by one-and-a-half to two times in Cheshire, Lancashire, Somerset or Cornwall, but by four to six times in Essex, Lincolnshire, Norfolk or Suffolk.[54]

It is no doubt true that farmers in the once great grain-growing districts were constrained, or inhibited, by traditions that placed grain growing at the heart of serious farming, and that they might have made greater changes and adjustments than they did. At the same time, they were more objectively constrained by their soils and climate, which made it difficult to contemplate the complete abandonment of grain growing and impossible to emulate the high degree of grass feeding of livestock in western districts. Hence, the increase in stocking ratios in the eastern districts, particularly of dairy cattle and to a lesser extent of beef animals, had to be sustained by growing feed grains and purchasing feedstuffs, especially oilcakes. Thus, it is likely that a disproportionate share of the national consumption of oilcakes, as well as of fertilizers, and a disproportionate amount of the farm machinery, were deployed in these eastern areas. It is perhaps not surprising that the effort and enterprise represented by these changes in production methods and end products were not more handsomely rewarded by greater growth in output and productivity, because they were characteristic of the areas most exposed to import competition and that tended to keep profit margins low. By the same token, enterprise and radical change were to be found in those districts and farming systems most directly challenged by competition while, in the more sheltered farming systems, change was largely a matter of judicious expansion of already well-established lines of business.

The overall picture, then, is that British agriculture between 1870 and 1914 was not supine, paralysed, and demoralized, meekly surrendering its home market to foreigners without putting up any resistance. The landowners were pushed into retreat, in rental incomes, in land values, and in power over farming, but they were not a dead-weight dragging British agriculture down. The farm workers either left the land, or were made redundant, in droves; but those that remained enjoyed rising real wages and, on average, became more skilled either with horses or with machines. The farmers bore the brunt of the crisis of the sudden collapse of prices. Many were badly buffeted, many were ruined. But, after a large turnover of tenants, it was the farmers who initiated and carried through the changes in methods and in the types and mixes of farm products that sustained not an agriculture in decline, but an agriculture of which its total product increased in volume and in value. Whether the farmers might have done more perhaps remains an open question. What is certain is that what they did depended to no small degree on the availability, and expansion, of manufacturing capacity to supply the farm machinery, the fertilizers, and the oilcakes that were essential elements in the response. These industries were not new in this period, nor did they grow to become 'great' industries of major importance to Britain's manufacturing economy. They depended, however, on making agricultural applications of techniques and processes initially developed for other purposes: the agricultural machinery industry was dependent on the engineering industry for its machine tools and for many of its components; the fertilizer industry was partly dependent on the chemical industry; and the oilcake industry on the oilseed-crushing industry. The nascent integration of agriculture and manufacturing industry was epitomized by this development of the agricultural and manufacturing industry was epitomized by this development of the agricultural suppliers, among whom both the agricultural machinery and the fertilizer industries became not insignificant export industries in their own right. The effect of using these manufactured supplies within farming was that agriculture became increasingly like any other industry, processing materials procured from outside and adding value to the finished products which were sold to the consumers. Thus, the transformations in British agriculture in these difficult years, accomplished in a free-trade regime and without subsidies from the taxpayer, pointed the way towards the later twentieth-century model of industrialized agriculture. Contemporaries certainly did not see the course of agricultural change between 1870 and 1914 as a success story: but contraction, relative or absolute, rarely is seen that way, and the collapse of a large part of Britain's cornfields made a particularly sorry sight. At the least, agriculture, as the first of Britain's major industries to go into decline, had coped with its decline with more flexibility and less long-term anguish and distress than other staple industries, such as coal, cotton or shipbuilding, which went down the same road later. At the best, a slimmer agriculture, by 1914, really was fitter, economically, than it had been in 1870: British consumers enjoyed cheap food, the cheapest in Europe, and British farmers and farm workers enjoyed higher real incomes.

Above all, agriculture had made a positive contribution to such overall economic growth as did occur, not just by releasing labour for more productive use outside farming, but by raising the productivity of the resources which were retained within agriculture.

Notes

1 M. J. Wiener, *English Culture and the Decline of the Industrial Spirit* (Cambridge, 1981) has been the most influential of these.

2 P. J. Perry, *British Farming in the Great Depression, 1870–1914* (Newton Abbot, 1974); H. Newby, *Country Life: A Social History of Rural England* (Totowa, New Jersey, 1987), p. 104.

3 S. Pollard, *Britain's Prime and Britain's Decline: the British Economy, 1870–1914* (1989), pp. 1–8 summarizes the most recent work on macro-economic growth rates, and Table 1.2 presents estimates of UK rates of growth in output per head.

4 The conclusion of the *Royal Commission on the Supply of Food and Raw Materials in Time of War* (B.P.P., 1905) XXXIX.

5 Quoted in A. Offer, *The First World War: An Agrarian Interpretation* (Oxford, 1989), p. 96.

6 H. Rider Haggard, *Rural England* (1906 edn), II, pp. 539, 541.

7 M. M. Postan, *An Economic History of Western Europe, 1945–64* (1967), pp. 57–60, 80, 84–5, 88.

8 Calculated from B. R. Mitchell, *European Historical Statistics, 1750–1970* (1975 edn), Tables C1 and K2. The differences between agriculture's share of GNP and share of labour-force illustrate its lower productivity rating.

9 Eurostat, *Basic Statistics of the Community* (Brussels, annual), e.g. 1978–9, Table of degrees of self-sufficiency; 1987–8, Table of Supply and Demand Balances.

10 P. E. Dewey, *British Agriculture in the First World War* (1989), Table 2.7, p. 16.

11 Gross agricultural output at current prices divided by numbers employed: for England, 1867–71, from T. W. Fletcher, 'The Great Depression of English Agriculture, 1873–96', *Economic History Review*, 2nd ser. XIII (1960–1), p. 433; for Britain, 1909–13, from Dewey, *British Agriculture*, p. 13. For the UK, C. H. Feinstein, *Statistical Tables of National Income, Expenditure and Output of the U.K., 1855–1965* (Cambridge, 1976), Table 54, supports a lower rate of growth, at constant factor costs, but derives from different, and superseded, estimates of gross agricultural output.

12 Offer, *Agrarian Interpretation*, chaps 7 and 8, especially pp. 94–5.

13 Offer, *Agrarian Interpretation*, pp. 107–10, 117–20.

14 R. J. Thompson, 'An Enquiry into the Rent of Agricultural Land in England and Wales during the Nineteenth Century' (1907), reprinted in W. E. Minchinton (ed.), *Essays in Agrarian History* (Newton Abbot, 1968), II, pp. 75–80: he estimated the average 'economic rent' per acre in 1900 as 4s.7d. out of a gross rent of 20s.

15 Estimates derived from A. Armstrong, *Farmworkers: A Social and Economic History, 1770–1980* (1988), pp. 120–1, 137–8; E. H. Hunt, *Regional Wage Variations in Britain, 1850–1914* (Oxford, 1973), pp. 62–4; and B. R. Mitchell, *British Historical Statistics* (Cambridge, 1988), Labour Force 25 and Prices 9.

16 Dewey, *British Agriculture*, pp. 62–3.

17 M.A.F.F., *A Century of Agricultural Statistics: Great Britain, 1866–1966* (1968), Tables 46–8: overall, cereal acreage declined by 25 per cent between 1870 and 1914, with different trends for the three chief grains.

Cereal crops in Britain in thousands of acres

	1870–4	*1890–4*	*1910–14*
Wheat	3558	2147	1853
Barley	2339	2086	1686
Oats	2691	3045	2964

18 *Century of Agricultural Statistics*, Table 42 (arable acreage), Table 70 (horses for agricultural use). A small proportion of field machinery was steam-powered but, although 600 sets of steam ploughing tackle were in use in 1910, the acreage ploughed annually was not large; Dewey, *British Agriculture*, p. 60.

19 Details are in F. M. L. Thompson, 'The agricultural, chemical and fertilizer industries', *The Agrarian History of England and Wales, 1850–1914*, VII (Cambridge, forthcoming).

20 Eurostat, *Basic Statistics* (23rd edn, 1984), Table of consumption of fertilizers in 1982–3. Figures for use of *Phosphatdünger* are for kilograms of 'pure nutrient content' of $P_2 O_3$ per hectare of total agriculturally utilized area: for GB 24 kg/ha (21.3 lb/acre) Superphosphate contained 25 to 27 per cent soluble phosphate, so that 21.3 lb represented 80–85 lb of superphosphate. Other countries, in 1982–3, used two to three times as much phosphate as Britain: W Germany 61 kg/ha; France 52 kg/ha or Belgium 71 kg/ha. In 1912–13 the 1.087 million farm horses contributed, at 12 tons per horse p.a., 13 million tons p.a. to the total supply of farmyard manure. Horse manure contained 0.61 per cent of phosphate, and the total horse-produced supply amounted to 5.5 lb/acre.

By the 1980s there was no farm-horse contribution to phosphate supplies.

21 Guano has been taken as containing 24 per cent phosphate, 17 per cent nitrogen, and 4 per cent potash; superphosphate and basic slag as 25 per cent phosphate; and Chilean nitrate and sulphate of ammonia as 20 per cent nitrogen. For the 1980s, Eurostat, *Basic Statistics, Stickstoffdünger* tables give N in kg/ha.

22 *26th Annual Report of the Chief Alkali Inspector* (B.P.P., 1890) XX, C. 6026, p. 12, commented that it was strange that British farmers did not buy sulphate of ammonia, because it was produced on their doorstep in every market town (by gasworks). *Departmental Committee on the Working of the Fertilizer and Feeding Stuffs Act, 1893* (B.P.P., 1905) XX, Cd 2372 and 2386, H. Voss stated that most sulphate of ammonia was used in manufacture of compounds, and little was used direct by farmers (q. 3205).

23 *Departmental Committee on the Adulteration of Fertilizers and Feeding Stuffs* (B.P.P., 1892) XXVI, C. 6742, evidence of H. Voss (q. 1868); and H. Voss in evidence to 1905 Committee (q. 3209). For 1913, Dewey, *British Agriculture*, p. 69. But *Report of Ministry of Reconstruction, Agricultural Policy Sub-Committee*, Part II (B.P.P., 1918) V, Cd. 9079, App. XIII, Potash, stated that entire pre-war supply of 23,000 tons p.a. was imported from Germany: if this meant 23,000 tons of K_2O, it was the equivalent of about 190,000 tons of kainite.

24 J. H. Clapham, *The Economic Development of France and Germany, 1815–1914* (4th edn, Cambridge, 1948), p. 213. The very high per capita consumption of potatoes in Germany also illustrated the gap between German and British living standards.

25 *Century of Agricultural Statistics*, Table 49 (potato acreages), Tables 51 and 52 (acreages of turnips and swedes, and mangolds). For predominant use of superphosphates on roots, with lesser use on cereals, *see* A. D. Hall, *A Pilgrimage of British Farming, 1910–12* (1914), pp. 4, 120, 126–7; *see also* Dewey, *British Agriculture*, p. 70.

26 *Century of Agricultural Statistics*, Tables 56–62.

27 This effect is mentioned in C. S. Orwin and E. H. Whetham, *History of British Agriculture, 1846–1914* (1964), pp. 282–3.

28 Total output equals gross farm income equals gross agricultural output equals value of all produce sold to the non-farm sector. Total factor income equals gross agricultural output minus cost of non-farm inputs (i.e., purchased feeds, fertilizers, machinery, and blacksmiths', farriers', and veterinary costs) equals income available to reward landowners, farmers and labourers. Factor income analysis is the most sensitive measure but, in the absence of data on non-farm inputs factor, shares are often expressed as shares of gross agricultural output:

(a) Rent, England and Wales, as percentage of gross agricultural output:

1872–3	35
1892–3	34
1909–13	20

Output from Fletcher, p. 433, and P. E. Dewey, 'Agricultural Labour Supply in England and Wales during the First World War,' *Econ. Hist. Rev.* 2nd ser. XXVIII (1975), p. 105.

(b) Factor shares of total factor income, UK, from Feinstein, *National Income* . . . Table 23:

	Rent	*Farmers*	*Labour*
1872–3	33	30	37
1892–3	36	21	43
1910–11	30	30	40

(c) Factor shares of total factor income, England and Wales, from Dewey (1975), p. 105:

1909–13	27	41	31

(d) Factor shares of total factor income, UK, from J. R. Bellerby, 'The Distribution of Farm Income in the U.K., 1867–1938,' (1953) reprinted in Minchinton (ed.), *Essays in Agrarian History*, II, pp. 261–78, especially Table 1:

	Net rent	*Farmers' income + Interest on capital*	*Labour*
1867–73	23	28 + 13	35
1884–96	23	24 + 13	40
1911–14	18	36 + 12	33

Bellerby, p. 270, shows an increase of 24 per cent in farmers' 'real incentive income' between 1867–73 and 1911–14.

29 Data are only available for UK as a whole: purchased inputs from E. M. Ojala, *Agriculture and Economic Progress* (1952), Table XIX; rent and labour from Feinstein, Table 23:

	Total costs	*Percentage of total costs*		
	£ million	*Rent*	*Labour*	*Inputs*
1870–6	179	31	34	35
1886–93	155	31	35	34
1911–13	177	24	33	42

30 *See above*, pp. 4–5 and note 13.

31 F. M. L. Thompson, 'An Anatomy of English Agriculture, 1870–1914,' in B. A. Holderness and M. Turner (eds), *Land, Labour and Agriculture, 1700–1920* (1991), p. 226.

32 Thompson, 'Anatomy', p. 234, for proportions of different crops sold off farms. Horse consumption taken as 1.4 tons of oats per horse p.a., equivalent to 2 acres of oats. Total acreages devoted to fodder crops, Great Britain, were, in '000 acres:

	1873	*1911*
Wheat	349	476
Barley	163	511
Oats	2007	2529
Potatoes	154	171
Turnips	2122	1563
Mangolds	326	452
Total for fodder	5121	5702
of which for horsefeed for farm horses	1884	2174
Total tillage, GB	13820	10475

33 Dewey, *British Agriculture*, p. 16. These import percentages, measured in calories, differ markedly in the case of meat (chilled or frozen, not 'fresh') from those in *Century of Agricultural Statistics*, Table 21, measured by weight:

Beef and veal	47.4% imports
Mutton and lamb	48.5% imports
Pigmeat	64.2% imports (this included bacon and lard)

34 Proportions calculated from Fletcher for 1867–71 and Dewey for 1909–13. In 1909–13 liquid milk accounted for 14 per cent of gross output: Fletcher does not have separate figures for liquid milk, butter, cheese, and cream, but D. Taylor, 'The English Dairy Industry, 1860–1930', *Econ. Hist. Rev.* 2nd ser. XXIX (1976), p. 590, estimates that in the early 1870s about 33 per cent of total milk production was sold as liquid milk, 25 per cent as butter, and 40 per cent as cheese; thus in 1867–71 liquid milk probably amounted to about 4 per cent of gross output.

35 Offer, *Agrarian Interpretation*, p. 94. Contemporaries also worried about the failure to compete with Danish, and later New Zealand, butter and cheese.

36 Taylor, 'Dairy Industry,' p. 590. *Century of Agricultural Statistics*, Table 22.

37 Taylor, 'Dairy Industry,' p. 592.

38 Calculations from Taylor, 'Dairy Industry,' Table 3 suggest that total gallonage of milk made into butter and cheese declined from 354 million

gallons in 1866–70 to 220 million gallons in 1911–13, while that consumed as liquid milk grew from 150 to 650 million gallons. In practice, some of the residual butter- and cheese-making was geared to using seasonal peaks in output when it could not all be sold liquid.

39 Taylor, 'Dairy Industry,' pp. 595–6; G. E. Fussell, *The English Dairy Farmer, 1500–1900* (1966), pp. 195–9.

40 Taylor, 'Dairy Industry,' pp. 597–9.

41 R. Perren, *The Meat Trade in Britain, 1840–1914* (1978), pp. 3, 116–17, 162–4.

42 Perren, *Meat Trade*, Table 1.1; Ojala, *Agriculture and Economic Progress*, Tables XII, XVI; Orwin and Whetham, *British Agriculture*, Tables on pp. 251, 267, 350. The total head of sheep remained stable in the grazing counties, 1875–1910, and declined by 25 per cent in the arable counties.

43 Perren, *Meat Trade*, p. 77 cites contemporary estimates of average carcass weight of 700 lb in 1839, and 730 lb in 1859. E. H. Whetham, *The Agrarian History of England and Wales, 1914–39*, VIII (Cambridge, 1978), p. 13, quotes 672 lb carcass weight for 1909–13 from the 1908 Census of Agricultural Output. If there was any real decline in average weight, which must be doubtful because of the possible inaccuracy of the earlier estimates, it could have been caused by earlier maturity of beef animals, produced by improvements in stock management and as a response to changes in consumer tastes.

44 In contrast to crop yields, the carcass weight ceiling has not been lifted in the twentieth century, although the downward drift in weights has been substantially a consequence of the demand for leaner meat and the tendency to slaughter at younger ages:

Carcass weights:	1925	628 lb
	1972	578 lb
	1979	589 lb

Whetham, *Agrarian History*, VIII, p. 182. Eurostat, *Basic Statistics*.

45 The detailed sums are more complicated, and convert all farm animals into livestock units with the formula: 1 horse = 1 cow = 7 sheep = 5 pigs. Mitchell, *Historical Statistics*, Tables Agriculture 6 and 7, show that numbers of livestock in Ireland, and in Britain, changed by similar percentages, 1867–1914. Ojala, *Agriculture and Economic Progress*, Table XIX, gives tonnages and values of feeds consumed in the UK. It has to be assumed that consumption per head of these feeds was the same in Ireland and in Britain, which overestimates for Ireland where a higher proportion of stock were grass fed; hence, the increases in consumption per head in Britain were probably greater than estimated in the text. Ojala overstates the total quantities of feeds by including the entire net import of maize, most of which was in fact eaten by non-farm horses (and some of which was distilled into the cheaper whiskies).

46 Calculated from quantities and costs in Ojala, Table XIX, and livestock numbers in Mitchell, Agriculture 6:

	Oilcake per head, p.a. horses + cattle	*Cost per head, p.a.*
1870–6	1 cwt	£0.53
1911–13	1.8 cwt	£0.61

47 *Century of Agricultural Statistics*, p. 60.

48 Ojala, Table XIX, estimated inputs for the UK as a whole, many of which were, or derived from, imports, which were not separately recorded for Ireland. Dewey, *British Agriculture*, App. J, estimated inputs for Britain for 1909–13; these can be back-projected to 1870–6 on the assumption that the GB/UK relationship remained constant at 83 per cent.

49 Output: Fletcher's estimate for England, 1867–71, scaled up by 20 per cent to measure GB; Dewey's estimate for GB, 1909–13, *British Agriculture*, pp. 13–14. Work-force: C. H. Lee, *British Regional Employment Statistics, 1841–1971* (Cambridge, 1979). Out of some 400,000 farmers, perhaps half were working farmers with family farms, using little or no hired labour.

50 Hired workers: Census Occupation Tables, 1871 and 1911 – all those, men and women, classed as shepherds, those working with horses and cattle, and other agricultural labourers, both indoor and outdoor. Output: Fletcher for England, 1867–71, and Dewey (1975) for England, 1909–13, scaled up to £140 million to allow for omitted products. Output per capita at current prices.

51 Details of methods of calculating county estimates in Thompson, 'Anatomy'. The decline in national output, 1873–1911, is not out of line with the decline of 11 per cent, 1870–6 to 1911–13, in UK gross output in Ojala, Table XVI. Thompson's national total is slightly higher, for 1873, than Fletcher's for 1867–71; and is much lower for 1911 than Dewey's for 1909–13, the main difference being in the value attributed to meat output.

52 With labour-force taken as hired labour + farmers' relatives: 2 to 40 per cent increases in output per head in Cheshire, Cornwall, Lancashire, Salop, Somerset, Staffordshire, Westmorland and the West Riding; 20 to 30 per cent decreases in Cambridgeshire, Essex, Herefordshire, Huntingdonshire, Lincolnshire, Norfolk, Suffolk, Surrey and the East and North Ridings. With the labour-force taken as hired labour + farmers' relatives + farmers, the two contrasting groups remain the same.

53 Real output per head measured for hired labour-force only. Wide definition of labour-force yields 0 to 11 per cent increase for the eastern group, 52 to 73 per cent increase for the western group.

54 Composition of total farm output, by counties, at current prices. The proportions are county relatives, not absolutes: e.g. in 1911 56 per cent of Cheshire output was from milk, but only 20 per cent of Lincolnshire output.

3

Apropos the Third Agricultural Revolution: How Productive was British Agriculture in the Long Boom, 1954–1973?

B. A. Holderness

The Third Agricultural Revolution?

In the twenty years between the end of food rationing in 1954 and British accession to the European Community in 1973, agriculture underwent a genuine transformation. This transformation turned upon a social reconstruction of rural life as least as far reaching as the institutional changes brought about by demographic pressure and the enclosure movement in what is still conventionally regarded as *The* Agricultural Revolution. The social changes had a direct bearing upon the achievement of greater productivity, although their consequences were perhaps not always positive. Social transformation was paralleled by scientific and technological improvement and this will form the principal theme of this paper. The results were sufficiently impressive to suggest that there occurred between 1940 and the 1980s what amounted to a *Third* Agricultural Revolution. As a phrase, it has not yet become established but it is a superficially fitting description of the cluster of developments signalled by a change of direction during World War II that touched the practice of agriculture and the increase of productivity in the following peace.

The very idea of a third agricultural *revolution* is, of course, suspect but, if we employ the word at all, even in the wayward sense favoured by historians, the mid-twentieth century should claim precedence over the past in the achievement of truly outstanding economic gains in output and productivity. There has been little opposition to Michael Thompson's description of the period 1815–80, as the Second Agricultural Revolution, driven as it was by diverse influences upon the increase of production, by guano and under-drainage as much as by rational estate management, not least because most contemporary writing was positive and even self-congratulatory.[1] But to what it succeeded is more problematical, because the notion of an Agricultural Revolution roughly conterminous with an Industrial Revolution in the 50 years before Waterloo no longer holds the stage in historical dramaturgy. The issue need not concern us in detail, but the lineaments of any previous metamorphosis are important to the historian who seeks to find in the latest episode of rapid acceleration common ground with past revolutions. The question whether growth depends upon profound transformation or is essentially the phenomenon of government engineering is relevant to the prognosis of long-term symptoms. James Caird performed a significant, but undervalued, service to mid-nineteenth-century agriculture by explaining the characteristics of sustainable improvement in the aftermath of Repeal.[2] There are plenty of rationalizers around today but no-one, I believe, has yet produced a convincing synthesis of their often random notions.

The general importance of change since the war is easy to state. Whereas between 1770 and 1850, for example, aggregate output from British agriculture probably doubled in real terms, when population trebled and self-sufficiency therefore actually declined, in the 40 years to 1975 real gross output at least trebled and self-sufficiency increased from about 40 to about 60 per cent in temperate products.[3] Physical productivity in the form of advanced yields or more efficient management of agricultural resources disclosed unprecedented gains. This was nowhere clearer than in aggregate improvements of output per worker because the labour-force halved between the 1930s and the 1970s. At the same time, the horse-power available for farming operations rose by perhaps tenfold. Farming in 1940 was not significantly different in structure and practice from farming in 1840. In 1980 the points of comparison with the past had become largely obliterated.

Nevertheless, despite the overt success of the agricultural sector after 1940, there was no reversal of the long trend towards marginalization of the industry in the whole economy. Thus, although the sector already contributed well below 10 per cent of Gross Domestic Product in the 1925 Census of Production and employed less than 10 per cent of the work-force, its position continued to shrink afterwards. By the mid-1980s, British agriculture produced only 2 per cent of GDP and employed about 2 per cent of the work-force of the island.[4] There is a 'downside' in this process but, because almost all economists assume that prosperity, expansion and economic growth depend upon a small and peripheral agricultural sector,

British experience ought to reinforce the point that populations which spend fewer resources on getting and consuming food have much greater potential for growth. The fact that the evidence for this is equivocal is to be explained in different but contingent terms. The easy gains in productivity from deserting agriculture for industrialization in Britain had been attained before 1890. There was no quick 'fix' available to the economy from redundancy in agriculture, because we speak of fewer than 750,000 individuals released for other work in the course of the present century. In that sense, the achievement of notably higher productivity in agriculture affected the British economy as a whole rather less than the similar process in France or Germany in the post-war period.

The essence of this third agricultural revolution lies in the envelopment of agriculture with science and technology. This was not an innovation of the mid-twentieth century; the recognition that experimental 'philosophy' could improve, or even transform, agriculture predates Robert Bakewell and Joseph Priestley. Nevertheless, even the contribution of Augustus Voelcker or Lyon Playfair, in nineteenth-century applied chemistry, or of Gregor Mendel, in early genetics, was much more peripheral to agricultural progress before World War I than the work of the John Innes Institute, Cranfield, Thames Ditton or the various research stations has been to development since the 1950s. Moreover, in agro-industrial technology, led by firms such as ICI, Fisons, Ford, Massey-Ferguson or ABM, the relationship to modern agriculture is much more direct than had been the case when Fisons was founded or when International Harvester and Ford first turned to producing tractors. After World War II, in effect after 1953, a mosaic of industries, producing commodities specifically for agriculture, came into being, to supply machines of unprecedented power and complexity, veterinary preparations, fertilizers, pesticides, improved seeds, fixed capital goods, together with the advisory or scientific facilities to support them.[5] Agricultural demand, not infrequently driven by public subsidies, was a prime mover in significant branches of industrial innovation. This is true no less of the processing of agricultural produce than of supplies to farmers. Mass production and mass distribution of primary commodities, such as milk, flour and sugar, had begun before 1940 and set a trend soon followed by others. In the 1950s and 1960s, they were joined by deep-freezing and the increasing standardization that transformed the processing of meat, bread, dairy produce and poultry.

There is, then, in modern times a reticulated, interdependent agro-industrial complex in which it is difficult to unravel domestic from international influences. But, for what my calculation is worth, I have attempted to estimate the contribution to the national economy of this interlocking set of activities. Between 1954 and 1972, i.e. during the period that was open to free trade in agricultural commodities, the processing of agricultural raw materials supplied from British farms, and the satisfaction of farmers' demand for capital goods, chemicals, seeds, etc., expanded somewhat more than agricultural production *per se*, by trebling in real terms. This wider contribution of agro-industry to Domestic *Net* Product in 1972 was approximately 12 per cent, although this does not reflect the farmers' share of demand for electricity, water, petroleum or their consumption of non-agricultural goods. Expressed differently, the results are similar: the share of agriculture in aggregate manufacturing output at least doubled in the 20 years before the oil crisis. But, even in 1972, the agricultural customers of these industries were too disaggregated to exert more than a modest influence upon their industrial strategies.[6]

The interpenetration of post-war agriculture and so many segments of industrial and commercial activity, not only reduces the significance of a simple sectoral analysis, but also piles Pelion upon Ossa, because we are confronted with an insecure, contradictory database upon which to construct estimates. Most of the wider questions I have posed are perhaps unanswerable, except in terms that would not have consoled Ixion. Agriculture is so flexible a term in contemporary usage that it may comprehend the whole domestic food industry and the idea of a residual, rusticated way of life. There is a sector specifically identified as *agriculture* (although it has frequently included forestry and fishing). Although I am not content to confine this appraisal merely to that sector, it is possible at least to produce a balanced discussion of changes in productivity and to interpret their influence upon the perception of agriculture within the political economy of the United Kingdom and its relations with the European Community only in terms of the agricultural sector. The positive features which have been well publicized by agricultural and by government interests were taken, even in the 1960s, to suggest that Britain was not, in economic terms, irretrievably backward and could perform as well as, or better than, overseas' competitors. This consolation was important psychologically; for most of the period after 1950, the British economy, but not agriculture, was beset with Job's comforters. The impression of success, and perhaps its accomplishment, was inspirational. The positive gains made by agriculture were undeniably substantial, but not much greater than were obtained in other agrarian systems of the long post-war boom. In Britain, however, they were seen as the justification of the *British* system of agricultural support and of the wider superiority of our farming, especially in contrast to the peasant societies of the near Continent.[7]

Trends in Productivity

The primary measure of productivity is output per worker which was transformed between 1951–3 and 1971–3. Hired labour in Britain declined by one-half in 20 years, from 800,000 to 390,000. For those holding land the position is less clear, because the statistics are surprisingly inexact, but it seems justifiable to reckon upon a fall of perhaps 10 per cent in numbers between these dates. Altogether, therefore, there was a decline in the agricultural labour-force of between 40 and 45 per cent, or from 1.2 million to 700,000. It is, however, not as simple as the figures suggest. Labour productivity is very complex, not least because the basis of calculation is inconsistent. If we assume for a moment that the work-force was not quite halved before 1973, what of output? In real terms, gross output in agriculture increased from £1,100 millions in 1951–3 to £2,451 millions in 1971–3, using a base for calculation in the first triennium of free trade, 1954–6, which means that the output per worker, irrespective of inflation, went up two-and-a-half times.[8]

Simple explanations of improved labour productivity need to be qualified by the structural analysis of employment and of the distribution of incomes within that employment. The figures just quoted are flawed, first because the basis of measurement changed significantly through time. From the late 1960s, the annual statistics excluded many of the smallest holdings in some key returns, but not in all. Thus, output data in 1951–3 and 1971–3 are broadly comparable but not the labour-force statistics. While the contribution of the excluded holdings was not great, the disappearance of certain enterprises from the statistics obviously distorts the measurement of aggregate labour productivity. We can resort to guesses but a different approach is more likely to cut a way through the thicket.

In the 1960s the Ministry devised a method of classifying farms by their employment characteristics. The 'standard man year' (S.M.Y.) of 275 days (later adjusted as a result of changed assumptions to 250 days) became a yardstick which could then be used to quantify labour units on the farm. This was a better definition than the distinction between full time and part time, because it was evident that many so-called full-time farmers were actually occupied for less than 275 days on their holdings. Conversely, because other surveys also indicated that many farmers really worked many more hours per day than the expected standard, we should take account of entrepreneurial commitment, although much of this extended work-time was the result of inefficient management. We can use the 1966 evidence of standard man years as a basis for backward extrapolation into the 1950s. It is rather easier to use S.M.Y.s than other later standards of employment, the British Size Unit, or the Agricultural Labour Unit, employed in the past few years upon criteria that cannot effectively be translated into past time. The A.L.U. is not even remotely congruent with the *contemporaneous* A.W.U. (Agricultural Work Unit) used by EC statisticians.[9] Accordingly, I have attempted to calculate the full-time equivalent labour-force in agriculture in S.M.Y.s. In 1966, after some adjustment to the data, the number of *standard men*, employees and employers alike, amounted to 525,000, who produced £4000 per head. The real decline in number, 1966–71, possibly amounted to 9 per cent in 1966-style S.M.Y.s. For 1951 the lack of detailed evidence of employment in comparable terms makes any estimate doubtful, but I *think* the equivalent number would be about 815,000 S.M.Y.s producing £1200 per head. On this calculation, employment in agriculture declined by 40 per cent between 1951 and 1971. Standardized per capita income rose by 275 per cent between 1953–5 and 1971–3 (triennial averages), by which time each man was producing almost £5100 per annum.

My second point is even more difficult to resolve. Labour productivity was also influenced by the different rates of decline for farmers and land-workers. The proportion of farm expenditure taken up by wages certainly declined in the 20 years but, in relation to farmers' entrepreneurial income, the wages bill actually fell pro rata by more than the simple decline in numbers. Thus, in 1953 wages seem to have accounted for 65 per cent of total labour expenditure, and farmers' entrepreneurial income for about 30 per cent. In 1972 the position was approximately reversed.[10] In other words, farmers' earnings (not to say their profits) rose notably faster than their workers' wages. Even on farms of comparable size, the farmers' average income rose by two-and-a-half times, 1953–72; allowing for the effects of amalgamation and the retirement of many elderly and inefficient landholders in the interval, farm incomes rose approximately fourfold in real terms, whereas wages barely doubled. I suspect this disparity is a normal condition in periods of rapid agricultural expansion and progress. It certainly fits the periods, 1770–1815 and 1840–75 reasonably well.

There are several corollaries arising from this situation. First, the decline in full-time wage employment was rather greater than in part-time employment. Secondly, the average number of hours per week in full-time employment declined more on paper than in reality, a fact not adequately included in the calculation. Thirdly, the average size of farms increased from about 70 acres (30 ha) in 1951 to about 120 acres (50 ha) in 1971. Fourthly, the differential improvement in entrepreneurial incomes has implications for capital formation, which may be important for the continuance of agricultural improvement, but it may equally result in an overcommitment to conspicuous consumption or investment, especially in the market for freehold land, because there is no reason to suppose that farm entrepreneurs have been more prone to plough back their profits into reproducible capital than other British business interests.

Ostensibly, the physical productivity of the land is easier to estimate than the productivity of labour. Farmers and governments not only measure the state of each year's harvest by the yields obtained but also compare themselves with neighbours and competitors. How much the soil would yield in return for the seed implanted was for centuries so much a matter of people's living or dying that the object of agricultural improvement was always expressed, until very recently, as the magnification of output. A fear of scarcity still animated the affluent West in the early 1970s, although true to say, the alarm of 1972–3 was exceptional in the long period that had elapsed since the Korean War ended. Plenty was taken for granted in the 1950s and 1960s wherever agricultural production was sufficiently advanced to draw benefit from scientific innovations and extensive commercial exploitation. It led to no serious attempt to restrain the impulse to raising productivity. Yet this plenty was easily won, or so it seemed to agricultural ministries striving to keep output and prices more or less in equilibrium in the interests of consumers and producers alike. Nevertheless, in the mid-century, British agriculture worked a wonder that has never been equalled. Between 1938–9 and 1945–7 gross output in agriculture doubled at current prices; it doubled again before 1954–5. By 1972–3 there had occurred a third doubling in output, and thereafter, not least because of inflation, it entered the Heaviside layer in the ten years down to 1983. In *real* terms, however, the real gains of the 1940s, at about 60 per cent, were not repeated until the 1980s. In the 20 years, 1954–73, real gross output rose by only 70 per cent. These are all impressive statistics, as any agricultural historian will know, but they require careful interpretation.[11]

There were substantial gains in yield across the board. In this respect, the period between 1953–4 and 1972–3 was notably more productive than the 1940s when most of the real gains were made from converting pasture to tillage. For several important products, average yields in about 1950 were only marginally higher than in 1885, when the first careful, wide-ranging survey was made. As a result, the increase in output *per acre* after the end of rationing is remarkable. The conversions of the war period greatly increased the calorific value of British agriculture. The energy output of agriculture in the 1940s at least doubled, as more potatoes, grain, sugar beet, and milk were produced at the expense of rotational grasses, pasture and meat. Because almost all the arable converted before 1947 remained in tillage in the '50s and '60s the volume increase in output then occurred in the achievement of higher yields.

The area planted with sugar beet and potatoes did not increase between the mid-1950s and the mid-1970s, whereas total output rose by 35 per cent and 20 per cent respectively from 1955 to 1972. Yields, in fact, increased by 35 per cent for sugar beet and 48 per cent for potatoes. Wheat output almost doubled, to reach 4.5 million tonnes in 1971–2. Barley almost quadrupled to 8.2 million tonnes. Both crops were more extensively cultivated in the 1960s, but yields also went up by 47 per cent and 36 per cent respectively. Such improvements of yield in 20 years or less were great by historical standards, but they have been totally eclipsed by further developments in the 1980s, in the case of wheat because of better seed, and in the case of barley partly because of the adoption of winter-sown varieties. These figures were naturally affected by the condition of the harvest from year to year, and also by decisions about cropping, taken usually in response to the government's Annual Review and Determination of Guarantees in time for sowing. There was, however, an unmistakable upward trend of output and of yields in the 1950s and 1960s which persuaded many 'experts' to believe that British efficiency would easily resolve the structural problems posed by accession to the Common Agricultural Policy (CAP).

Efficiency was even more evident in the meat trade. Production of meat for the market increased from 1.73 million tonnes in 1957–8 to 2.93 million tonnes in 1973–4. This occurred even though the number of livestock other than poultry increased only modestly. Improved breeding, a quicker turnover in stock keeping and better feeding contributed to this change. What makes translation of net output figures into live animals difficult is the relative switch in preference from beef- and mutton-producing stock to poultry and swine. In effect, of course, the popularity of poultry affected total meat production only marginally. The great success story of the period was the mass production of broiler fowls, but even a tenfold increase in their number, when each bird averaged only 4 lb (1.8 kg), could never have offset a major fall in the number of beef cattle. But, after the war, the beef trade revived, so much so indeed that, even without the contribution of poultry, meat output still doubled between the early 1950s and the early 1970s. Other aspects of stock farming displayed the same upward trend. The average yield per dairy cow rose from about 690 gallons (3070 litres) to 860 gallons (3827 litres), which understates the real gains in productivity made since the period of austerity in the 1940s. Egg yields similarly went up from under 150 per bird in a year to 220 per bird.

Much of this raised productivity must be attributed to the interpenetration of industry and agriculture. Without purchased fertilizers and feeding stuffs, without lime or basic slag, without the benefits of mechanization and genetic manipulation, the figures cited would not have been attained and were probably unattainable. The contrast between organic (i.e. non-interventionary) farming and scientific or high-investment farming is evident enough, not least in the higher prices charged for the former kind of produce. There is, however, an insufficient body of statistical data to measure the differences accurately. In the past few years, many farmers on various soils have gained first-hand experience of organic and of scientific practices. There has always been a market for organic produce but, on many farms in the 1950s and 1960s, growers who refused chemicals and denied themselves the services of heavy modern machinery were not only few but held to be rather eccentric. It is difficult to obtain a satisfactory account of their yields and the costs at which they produced their crops, or indeed how their profits measured up against those of conventional capitalist producers. Direct evidence from several farmers of my acquaintance who have grown familiar with organic husbandry convinces me that it generally produces yields between one-third and one-half as great as scientific farming. Thus, my informants disclose mean yields for potatoes and wheat on organically farmed land of 19 tonnes per hectare and 2–3 tonnes per hectare against 40 tonnes and 6–8 tonnes per hectare on the same farms, 1985–7. Costs overall were not dissimilar, that is to say that there were no easy savings to be made from giving up chemicals. One correspondent who grows 40 acres (16 ha) of wheat organically and 400 acres (160 ha) scientifically in most years observed that he could produce no more of the former because his holding was not equipped for traditional systems of husbandry and he and his workers had either forgotten or had never known how to manage rotations successfully. What justifies the organic producer in the 1980s is the notably higher prices (up to twice the level of conventional produce) he or she can expect for organic stuff. Given the present faddism, this premium is understandable, but it certainly did not obtain in the 1950s and 1960s. Yields of all organically produced crops probably increased after the war, because of better seed, but there can be no doubt that the increase was modest, and the difference in productivity still favoured the scientific farmer by a factor of two in the 1960s. That may well be the most comprehensible gauge of achievement for the third agricultural revolution.

The idea of a transformation of British agriculture at the mid-century is therefore justified by these impressive figures, but there is a context in which they should be read that qualifies their impact. First, the progress in raising productivity was not unique. Most developed temperate agricultural regions displayed similar improvements. Secondly, although much was made of the special efficiency of British agriculture in the context of Western Europe, there is no doubt that equal or superior gains accrued to some rural economies dominated by much smaller farms. Thirdly, the costs, and not merely the ecological drawbacks of agri-business, made the growth achieved in this country appear increasingly vulnerable. The phrase, 'the limits to growth, or to expansion' entered the vocabulary of rural economists during the 1960s. In a sense, indeed, the debate in its economic aspect echoed the mid- nineteenth century argument about the benefits of high farming. One point was novel, that these costs were often not borne by those who had a direct interest in managing the land.

International comparisons tended to express particular prejudices and to ignore inconvenient data. Broadly speaking, the British view, even in the context of negotiations to join the CAP, was that the logic of mass production in agriculture, of economies of scale, was unassailable. Peasant regimes were intrinsically inefficient; they could provide neither plentiful nor cheap foodstuffs to satisfy urban, industrial societies, which therefore were required to pay a premium on the consumption of produce from a backward agriculture, supported in its ineptitude by high tariffs and/or a producer-oriented CAP. British progress in raising productivity was therefore self-justifying; it betokened entrepreneurial flexibility and the fruitful copartnership of market-sensitive farmers and a liberal government. Even the model of Denmark, so much used in pre-war prescriptions of agrarian reform, was effectively discarded, because that country seemed to offer no example worth following by agriculturists protected with subsidies and convinced by the economics of large-scale enterprises. Eric Nash, otherwise sceptical of much agro-economic orthodoxy in the 1950s, believed that the best agricultural regime was one which employed the least labour in relation to the total economy. Because Britain had for the first half of the century already found employment in agriculture for a notably smaller percentage of the work-force than even the Netherlands or Denmark, the case was made. The fallacy is not difficult to spot. Agriculture in Britain had become marginal to the economy by 1925 chiefly owing to the preference for free trade and our dependence upon overseas' supplies. Moreover, the great spurt that occurred after 1940 in agricultural output, while it effectively reduced this dependence upon imports, actually built upon levels of productivity that were already high by international standards. British agriculture, through prosperity, depression and prosperity, 1850–1950, yielded output per acre greater than the New World, France or Germany. But non-labour inputs were almost invariably higher, so that net product per acre/hectare, showed a differential of out-turn in various regimes less pronounced than physical output alone implies.

In cereal production and in dairying in the 1920s, gross physical productivity was about one-third greater in Britain than in northern France, whereas net product was probably about one-fifth higher. These differences, moreover, were more-or-less eliminated by the 1960s. If we compare cereal yields in the Beauce or Picardy with those of south-eastern England, 1972–6, the difference was negligible. Norman dairy farms also produced yields of liquid milk not far short of those in Somerset. This convergence of output statistics from similar enterprises in Western Europe, particularly in the 1950s and 1960s, implies a larger rate of improvement on the Continent than in Britain. The trend was already discernible before the war, although it was especially marked only on select, capital-rich holdings. By contrast, the small or medium-sized farms of Flanders, the Netherlands and Denmark already exceeded best-practice yields in Britain by 1920. These farms, of course, made inefficient use of labour: peasants, even when they used spade husbandry and intensive horticultural rotations, were traditionally underemployed. Yet, even in the 1950s, British agriculture concealed a considerable degree of underemployment, as we have demonstrated. The differences in *physical* productivity persisted for, by 1970, Denmark, the Netherlands, Flanders and South Brabant all produced more per hectare of most agricultural commodities than Britain.[12] In strict truth, one should perhaps describe a region around the North Sea, from Lothian and Jutland to Picardy and the Lower Rhine, in which, by the time the CAP embraced the whole, already enjoyed a high and rising level of productivity, irrespective of rural social organization.

My point is that British experience after the war was not unique. The third agricultural revolution encompassed all temperate latitudes outside the Soviet bloc. It took place in spite of institutional arrangements, under Protection, Deficiency Payments and Free Trade, in inward-looking and in world-centred agrarian economies. This, given the impetus imparted by scientific innovations and marketing initiatives, is not surprising. New ideas quickly became common currency so far as progressive, adaptative farmers were concerned. There were nevertheless important differences between agricultural regions and nationalities in the rate of adoption of, or in preference for, new scientific and commercial opportunities. Mechanization was not simply a process of cultural diffusion, by which I mean that a common stock of inventions gradually spread out into various agricultural regimes. For a peasant, for example, the tractor is a direct and specific replacement for the horse or ox, whereas for farmers in Norfolk, or Nebraska, or Picardy, tractors represent savings in labour, speed of cultivation and adaptability in using an indispensable range of implements, all of which in their turn deliver similar results. There is a very wide difference of scope, utility and function between a tractor delivering 150 bhp and one with a torque equal perhaps to two real horses. In different economies, mechanization as a process, approved because of its modernity, produces signally different results in terms of capitalization and in terms of agricultural practice. Britain in 1970 was among the most highly and completely mechanized of the world's farming systems. There were fewer than 18,000 horses on farms and over 480,000 tractors. Nearly 40 per cent of gross fixed capital formation of about £300 million was spent on plant and machinery, which required about £200 million a year in maintenance expenses. In constant prices, the zenith of gross fixed investment in agriculture was reached in 1974 at about twice the relative level of 20 years previously.[13]

Even before the Yom Kippur War, people had begun to question the wisdom of extensive dependence upon such elaborate and often under-employed machines. The practitioners of high-capitalist farming in Britain were accused of being prodigal in the use of mechanical energy.[14] During the oil crisis it also seemed that British agriculture had become vulnerable to forces beyond the farmer's control. As it was, as Kenneth Blaxter pointed out in the mid-1970s, agricultural prosperity, the whole basis of the recent revolution, had been made precarious by an overcommitment to mechanical energy. Except so far as mechanization reduced the versatility of skilled workers and therefore made retreat into less highly exposed ways difficult, the case against machines is unsustainable, outside the particular context of the oil crisis. In the 1980s, an assault upon a different flank has been mounted, because ecologists have decided that the burning of hydrocarbon fuels is not so much expensive as antisocial – an opinion prefigured in the debates of the 1960s about the deficiencies of high farming. On the question of fuel costs, however, the argument is confuted by the negligible share of expenditure upon all fuels in farmers' accounts throughout the 1950s and 1960s and again in the 1980s. Moreover, machinery costs have seldom exceeded 20 per cent of total expenditure.

The real question, however, is not answered by statistical data. The size of machines and energy output greatly increased after 1950. So the question becomes, did farmers need all the firepower they deployed? My own computation suggests that agriculture could call upon horsepower in 1972 ten times greater than in the inter-war years.[15] The costs of servicing this power altogether increased through time even though, for a period in the late 1950s, the retail price of tractors, combine harvesters and milking machines fell in real terms. There is, I suppose, an element of creative accounting in the factorized expenditure tables, especially in the 1960s, not least because farmers were still maintaining a depreciated stock of implements for use as spares or in emergency. The most we can say is that the clear dichotomy between labour and capital in farm operations and accountancy hardly existed.

One explanation of the farmers' plethora of machinery in the 1960s is that their labour supply in itself had become precarious owing to the flight from the land. So far as I can tell, the exodus of labour from agriculture was less the result, than the cause, of mechanization. For example, some fenland farmers, who grew potatoes and other vegetables, invested in harvesting machines *only* after they had found the traditional gang system had become unreliable. In strict logic, the benefits of mechanization could be demonstrated in accounting terms, but inertia, due in part to an attachment to social convention (the reluctance to replace a system of labour that occupied women and the poor) and in part to the belief common among farmers that buying a machine employed a few weeks in the year was a fool's game, delayed action until necessity forced it upon the landholder. Machines offered one other major advantage hardly any more clearly understood, for, by supplanting horses, tractors released land for commercial exploitation. In the context of large-scale agriculture, this was not perhaps a vital consideration, but on peasant farms it produced far-from-negligible results, which may give another insight into the potential advantages of *intensive* peasant cultivation. The machine is ancillary to the person, not the person dependent upon the machine.

Effects of Subsidies

It was an essential part of government policy, enshrined in the 1947 Act, to support agricultural prices. This policy, first broached in the 1930s as a means of relieving depression, was pursued vigorously after the war when the imperative necessity was to increase output to make import savings. The method adopted after 1947 was to follow the wartime practice of direct purchase and direct distribution. So long as rationing lasted, there was no obstacle to price fixing by government. In the 1950s the problem changed. First, the reappearance of plenty in world food markets and the consequent downward pressure upon international prices caused conflict in Britain between the Treasury and the Ministry of Agriculture, Fisheries, and Food (MAFF). Secondly, because there was a threat to farm incomes from the general commitment to free trade in foods in this country, the Ministry devised an extensive system of deficiency payments, that is to say, of subsidies, to compensate for lower international prices.[16] In effect, there was a two-tier system in Britain – one the prevailing price for a product, and the other, a price determined by the Minister to satisfy the farmers. Actually world prices were seldom current in this country despite the general policy of 'cheap food', but the differences between producer and consumer prices were still considerable in the 1950s and 1960s.

This gives us a means of measuring the effects of subsidies. Eric Nash and Gavin McCrone produced rather sceptical analyses of the national benefits to accrue from subsidizing agriculture both from consideration of actual government expenditure and from comparison of real and protected prices.[17] There is little doubt that, by 1960, the United Kingdom had one of the most highly protected agricultural industries in the developed world. Acknowledgement of this led not to a significant decline in annual support but to a change of emphasis away from open-ended deficiency payments towards production or improvement grants. In due course, however, these reinforced the original problem. During the 1960s governments were unable to slough off their commitment to supporting producer prices although, from 1962, a ceiling was imposed by introducing the concept of 'standard quantities', above which no support would be offered.

Official subsidies of all kinds varied between about £210 and £280 millions per annum, 1955–71, with a long-run average of £250 million. Through time, the burden fell, both as a result of inflation and in terms of total public expenditure.[18] The proportion of agricultural gross output contributed by direct subsidies obviously declined in the same period but, because there were always hidden or indirect subsidies propping up sections of the industry, especially in dairying and beet-sugar production, the rate of decline was not considerable. In the middle 1950s, Eric Nash calculated that the total public support of agriculture was about £350 million. The £90 million he proposed as the level of hidden subsidy which, in itself was probably an understatement, subsequently grew faster than the official subsidies.[19] By 1970–1/1972–3, the true level of subsidy probably averaged £420 million to £430 million against £365 million in 1955–6/1959–60. As a proportion of gross product from agriculture, support fell significantly from 24 per cent in the 1950s to 16 per cent in the early 1970s. In comparison with international, i.e. European import prices, the level reached 32 per cent of gross output in the late 1950s and then declined to 23 per cent. According to McCrone, Britain's support of agriculture around 1955 was higher than any Western European country except Switzerland (30 per cent) and Finland (42 per cent) and exceeded the subsidization of the industry in France, West Germany and Italy by about one-third (24 per cent against 15.5 per cent).

In the context of the policy to improve British agriculture, the high cost of subsidies was a price most experts were willing to pay, at least in the 1950s. Because import penetration into Britain did decline in the period, with self-sufficiency rising from under a half (45 per cent) of all temperate products in 1954–7 to two-thirds in 1970–2 (66.9 per cent), the policy could be judged successful. Much ink was spent in drawing up schemes to make use of agricultural improvement to reduce the deficit on the balance of payments. It became a cornerstone of Austin Robinson's applied economics in the 1950s and reappeared as a serious proposition in the later 1960s under the influence of Asher Winegarten of the National Farmers' Union (NFU).[20] The problem, as ever, was to evaluate the benefits accruing to the national interest against the costs of effective producer protection. Grants to aid production or to promote improvements were certainly seen as a means of raising efficiency in the magnification of production but, too often, their introduction was determined by *political* necessity. So far as their effect was to increase output of commodities, such as barley and milk, that were in surplus in the 1960s, improvement grants were a double-edged weapon.

More damagingly, the argument that subsidies underpinned the strategy of rising productivity was overthrown by comparison with Dutch and Danish experience, where prices closely resembled current world prices for relevant commodities and little direct subsidization was in place after the war. Yet the yields and profits of farmers in those countries at least equalled those of their British counterparts. Farm net income in the UK in 1956–8 averaged 21.5 per cent of gross output against 26 per cent and 23 per cent respectively in Denmark and the Netherlands.[21]

Nevertheless, the general effectiveness of subsidies in raising net farm income in Britain is irrefutable. Thus, farm income in 1936–8 was about one-seventh of gross output (£41 million out of £282 million) against 21.5 per cent in 1956–8 (£326 million out of £1508 million) and 21.4 per cent in 1970–2 (£577 million out of £2690 million). In comparing levels of subsidy with net income, we have already seen that the increase must have owed very much to government support. If we recalculate the income data on the basis of 'European import prices' in the 1950s, for example, the ratio declines to 18 per cent of the (adjusted) gross output. The windfall was treated in the post-war boom as a *sine qua non* by the farming lobby. Any attempt to curtail or dismantle the system was greeted with howls of complaint. It was, nevertheless, the opinion of Eric Nash, embattled against Austin Robinson and his co-adjutors, that the *institution* of a regular system of support distorted the process of adjustment in agriculture *vis-à-vis* the structure of international trade.[22] This notion was politically reprehensible to farmers' leaders brought up with the myth of the Great Betrayal in 1920. Moreover, by the late 1960s, the Nash view had fallen out of fashion because the new imperative of agricultural policy was integration into the CAP at some appropriate time. Opposition to agricultural subsidization was mostly much less coherent. There had always been a band of politicians and their allies who complained that support for the industry was wasteful. In its implementation, subsidization tended inevitably to favour the large-scale producers, including some at least of the processors and intermediaries of the food trade, as it did in France, Italy and the United States, partly because the better-off knew how to manipulate the devices of government, but chiefly because subsidies and grants were related directly to quantity in gross, and only a little of the public benefaction was bestowed upon the smaller, inefficient grower or stock keeper. In the period of direct subsidization, 1933 to 1972, the number of agricultural enterprises (i.e. not simply holdings) fell by about 25 per cent, and the average size of holdings in Britain more than doubled. Furthermore, given the relationship of subsidies to farmers' income, their virtual dependability may well have encouraged agriculturists into extensive land purchases. Transfers of public money intended to secure the food supply were used to enrich a vociferous and prospering minority-interest group. Farmers, who before the war enjoyed incomes that placed them approximately alongside engine drivers or ironfounders, had, by the late 1950s, passed clearly into the middle class with average incomes equal to university lecturers and salaried accountants or solicitors.

One of the most inconvenient statistics of the third agricultural revolution is the inexorable increase in the proportion of agricultural land in this island owned by the occupiers. This is not a unique trend; indeed, what was odd about British agriculture around 1900 was the high proportion (about 85 per cent) of rented land. The problem in the period since 1945 has been the conspicuous consumption involved in buying freehold farm land. In 30 years before 1975 the percentage of cultivated land that was owner-occupied increased from 30 to 60 per cent, that is to say, about 4 million hectares of land under crops and grass changed hands from *rentiers* to occupiers. Despite the apparent plenty of freehold property available, the price of land with vacant possession increased notably faster than the general index of prices:

Table 3.1 Average price of agriculture land in Britain 1951/5–1970/3

	average price of agricultural land	*index*	*agricultural price index (based in 1955)*
1951/5	£193.5 per ha	100	94.8
1961/5	£432.4 per ha	223	97.4
1970/3	£1149.0 per ha	594	126.1

Source: Oxford Institute of Agricultural Economics land price series

Land hunger is a normal complement of prosperity and, in the twentieth century, it has been reinforced by the dissolution of the great estates because many tenants were pressed to buy the freehold of their holdings for fear of the alternative. But, in the 1960s, a new element was present, the rage for amalgamation and, with it, the rivalry of neighbours for land, which led to an immense increase in farm mortgage debt. Total interest charges upon agriculture of which, by the late 1960s, at least 70 per cent was owed on mortgages, increased from £14 million per annum 1951–5 to £22 million in 1961–5 and £55 million in 1970–3. The promise of protection gave farmers the confidence to enter the land market and the fact of subsidization fructified their ambition. There is a curious irony of the later 1960s that the annual sums invested in land approximately equalled the total of public subsidies to agriculture.

Notes

1 F. M. L. Thompson, 'The Second Agricultural Revolution, 1815–80'. *Econ. H. R.* XXI, 1968, pp. 62–77: *See also* chapter II in this volume. No agreement about the description of particular periods in agricultural history exists: C. S. Orwin, for example, called the transformation between 1760 and 1820 the *second* agricultural revolution, after the

Neolithic discovery of agriculture; cf. D. B. Grigg, *English Agriculture: An Historical Perspective* (Oxford, 1989). Also M. Tracy, *Agriculture in Western Europe* (2nd ed. 1981).

2 James Caird, *High Farming, under Liberal Covenants* (1849); *idem*, *Our Daily Food* (1868). Cf. E. L. Jones, 'The Changing Basis of English Agricultural Prosperity, 1853–73': E. L. Jones (ed), *Agriculture and the Industrial Revolution* (Oxford, 1974), pp. 191–210. Cf. C. A. S. Reading, *The Efficiency of British Agriculture* (7th report, 1980).

3 B. A. Holderness, 'Prices, Output and Productivity', *Agrarian History of England-Wales, Vol. VI, 1750–1850*, pp. 74–78. Grigg, *English Agriculture*, pp. 4–15; HMSO *Annual Review and Determination of Guarantees* (after 1973, *Annual Review of Agriculture*); G. McCrone, *The Economics of Subsidising Agriculture* (1962) Ch. 3; R. Mordue and J. Parrett, 'United Kingdom self-sufficiency in food, 1970–78', *Economic Trends* 312, 1979, pp. 151–5.

4 CSO, *United Kingdom National Accounts*, digested in the *Annual Abstract of Statistics. See also*, MAFF, *A Century of Agricultural Statistics 1866–1966* (1968); D. K. Britton (ed), *A Hundred Years of British Food and Farming: A statistical survey* (by H. F. Marks), (1989).

5 *See* e.g. B. A. Holderness, *British Agriculture since 1945* (Manchester, 1985) *passim*; A. Edwards and A. J. Rogers (eds), *Agricultural Resources* (1974).

6 Based on the pioneering statistical explorations of E. M. Ojala, *Agriculture and Economic Progress* (Oxford, 1952) and carried forward by the present writer using virtually the same series of data.

7 E. F. Nash, *Agricultural Policy in Britain: Selected Papers* (Cardiff, 1965); G. Halle. 'The Economic Position of British Agriculture', *Econ. J.* (1959), both essentially sceptical.

8 MAFF, *Annual Agricultural Statistics of England and Wales*; *idem of Scotland* (HMSO, annual series).

9 G. H. Peters, 'Capital and Labour in British Agriculture . . .' *Farm Economist* 11, 1967; Econ. Dev. Committee for Agriculture, *Symposium on Agricultural Manpower* (NEDO, 1969). The A.L.U. (MAFF, *Agricultural Labour in England and Wales*, 1978 return) and the A.W.U. (Eurostat: Agriculture: Statistical Yearbook, 1980 ed.) are *theoretically* the same but produced data in 1975 of 575,000 and 625,000 units respectively. In general, R. Gasson, 'Labour', in Edwards and Rogers, *Agricultural Resources*, is invaluable.

10 MAFF, *A Century of Agricultural Statistics*; Britton, *A Hundred Years of British Food and Farming*, App. 4, 7, especially Table 4.11, pp. 142–3.

11 *Ibidem* App. 7; MAFF Departmental Net Income Calculation.

12 The chief source is the Eurostat, *Statistical Yearbook for Agriculture*, 1960 and 1975 eds; INSEE, Paris, *Statistique Regionale de la Normandie* and *S. R. de la Picardie, Artois* (annual but 1979 used).

13 *Ibid*, App. 28, 29; *Annual Agricultural Statistics of the U.K., passim*; B. Hill, 'Capital' in Edwards and Rogers, *Agricultural Resources*.
14 *See* M. B. Green, *Eating Oil: Energy Use in Food Production* (Boulder, Colorado, 1978) Ch. 2; G. Leach, *Energy and Food Production* (Guildford, 1976).
15 This is surprisingly difficult to estimate because the available data tend to deal in numbers rather than capacity. I have used the reports in the journal *Power Farming* as a gauge of capacity, and distributed the quantitative data from the Annual Statistics according to the criteria implied in the technical press.
16 *See* especially P. Self and H. J. Storing, *The State and the Farmer* (1962) Ch. I, III, X; G. Hallett, *The Economics of Agricultural Policy* (Oxford, 1968) Ch. 10.
17 E. Nash, *Agricultural Policy in Britain, idem*, 'the Competitive Position of British Agriculture', *J. Ag. Econ.*, XI, 1955; G. McCrone, *The Economics of Subsidising Agriculture*, Part II.
18 HMSO, *Annual Review and Determination of Guarantees, passim*; Britton, (ed.), *A hundred Years of British Food and Farming*, Table 8.1, p. 153.
19 Nash, *art. cit.*, *J. Ag. Econ.*; McCrone, *The Economics of Subsidising Agriculture*, pp. 48–50.
20 E. A. G. Robinson and R. L. Marris, 'The Use of Home Resources to Save Imports', *Econ. J.*, 1950; Robinson, 'The Cost of Agricultural Import Saving', *Three Banks Review* (1958); L. Moore and G. H. Peters, 'Agriculture's Balance of Payments Contribution', *Westminster Bank Review* (1965); A. Winegarten, 'Agriculture and the Case for Import Saving', *Hill Samuel Occasional Paper, 5*, (1970).
21 Based on comparison of national statistics for each country, but *see also*, FAO, *Output Expenses and Income of Agriculture in some European Countries*, 3rd and 4th reports; K. Skovgaard, *The Pattern of an Unsubsidised Agriculture* (Wye College, 1960); E. F. Nash and E. A. Attwood, *The Agricultural Policies of Britain and Denmark* (1955).
22 See, e.g. McCrone, *The Economics of Subsiding Agriculture*, Ch. 2, 5.

4

Railways and the Development of Agricultural Markets in France: Opportunity and Crisis, *c.* 1840–*c.* 1914

Roger Price

Introduction

Ancien régime bureaucrats, faced with the disorders caused by the alternation of abundant harvests with penury, and with the substantial fluctuations in prices over time and between places, were attracted to liberal economic theories. These postulated that a free market was most conducive to the adjustment of supply and demand, and to the encouragement of increased production. The theories failed, however, to take adequate account of the shortcomings of the pre-industrial marketing system, caused by poor communications, and the resultant slowness and high cost of transporting bulky commodities, together with the imperfect diffusion of market-place information. Considerable investments in the improvement of waterway and road networks were therefore made by governments anxious to stimulate economic activity. This was to be a key factor in the acceleration of the process of industrialization and of the commercialization of agriculture evident from the second half of the eighteenth century. In spite of this, it can be argued that the basic features of the pre-industrial communications system survived until the railway age. The second half of the nineteenth century saw a far more substantial improvement of existing forms of transport than had ever previously occurred, as well as the injection of a whole new transportation technology into the economy.

Evidence to support this assertion is not difficult to find. Most notably, there were few major rivers capable of mass transport, on a regular basis, over long distances. The canals constructed to connect these had proved to be costly to construct and inefficient in operation. Poor maintenance and horse traction imposed severe limits on movement by road. Thus, although the basic configuration of the modern transport network, linking the major population centres, was established, its quality and carrying capacity remained limited. The major routes across the plains, along river valleys or by coastal shipping were poorly connected with their hinterlands and imposed an essentially linear pattern on the movement of traffic. This inevitably restricted their effectiveness as stimuli to economic activity. High transport costs, loss of time and the uncertainties of activity in markets distant in terms of space-time inevitably engendered caution among business people. The high cost of access had the result of reducing the effectiveness of the overall marketing system. Much of the countryside in particular remained isolated, with farmers committed above all to securing family subsistence and hoarding money as a source of security and means of purchasing land rather than commodities. Lepetit concludes that only the north-east, with its relatively effective waterways and a dense road network, stands out from this picture of otherwise general mediocrity, an obvious instance of the impact of natural conditions upon market opportunities.[1]

Assessing the impact of continuing investment in the improvement of communications during the eighteenth century and first half of the nineteenth century in a precise fashion remains virtually impossible in spite of the information contained in contemporary enquiries and traffic censuses. Their lack of homogeneity makes impossible statistical comparison of the effectiveness of the transport network over time. What is clear from the declining spatial variation in cereal prices is that the centuries-long process of market integration was accelerating. Moreover, the substantial increase in agricultural production is evidence of the ability of farmers to respond to changing market conditions. The balance between population and food supply remained precarious, however, with low-yield agricultural systems susceptible to frequent climatically induced harvest shortfalls. The information on cereal prices also suggests that, in spite of growing integration, a coherent national market was far from having been created. It reveals the existence of major isolated regions including the uplands and much of Provence. This is not to say that there was an absence of inter-regional trade, far from it, but that the shortcomings of the communications network overwhelmingly oriented actors in the market-place towards local and regional activity, and that distinctive marketing regions remained only loosely connected.

Following this brief introduction, I want to examine the impact of railway construction, from the 1840s, upon agriculture; to consider the interrelationships between the developing transport networks and the economic structures they connected as a prelude to an analysis of the flows of traffic generated; and finally consideration of farmers' reactions to changing market conditions. The essential question is, then, how did the economic actors respond to the concurrent injection into the economy of a new transport technology (the railway), and of a new means for the transmission of market-place information (the electric telegraph)?

Railway Construction

The improvement of communications clearly meant an increase in the social capacity for transport due both to the new potential for bulk carriage and the increased rapidity of movement. Reduction in the cost of transport had the effect of changing cost-distance equations, with important effects on the structure of the market for agricultural produce, and ultimately upon the spatial organization of farming. In considering the effects of railway construction on agricultural markets, however, it must be stressed that this was only one among a number of factors promoting change, and itself needs to be considered as part of a cluster of innovations affecting transport. One needs also to bear in mind the American economic historian Fogel's warnings against exaggerating the significance of railway development. It would certainly be excessive to claim that the railways were in all circumstances an essential prerequisite for the integration of national and international markets. Moreover, virtually all farmers were already to some degree involved in commercial activity. Further improvement of road and waterway networks would anyway have provided an additional stimulus. Nevertheless, railway development in France did more than promote a transfer from pre-existing forms of transport; and its impact on the economy was more substantial than in the case, for example, of Britain, described by Patrick O'Brien, in *The New Economic History of the Railways*,[2] as 'a small country with integrated commodity markets adequately served by water-borne transport into the major centres of population'. In a place and period in which the elasticity of demand for transport was high, railways were to be the main cause for a substantial increase in agricultural traffic. This in part resulted from the release of already existing regional surpluses of cereals and other produce, and also because the provision of improved access to markets constituted a compelling incentive to increase production. Moreover, the improvement in transport facilities did not represent a once-and-for-all innovation. Stimulated by a lengthy period of constantly growing demand, it would be extended spatially and also subject to technical improvement further to improve efficiency.

Modernization of the market structure had three other essential bases. These were: population growth; increased urban demand; and the contraction of self-sufficiency in the countryside. Demographic statistics give some indication of the growing importance of urban markets with the percentage of total population classified as urban rising from 25.5 per cent in 1851 to 44.2 per cent in 1911. The rate of urbanization and the development of market incentives was slow in comparison with some other Western European societies but nevertheless represented redistribution of population on a substantial scale. A smaller proportion of an increasing farm product was required for local consumption, and a far larger share needed to be commercialized.

The supply of urban markets is obviously a basic determinant of the structure of agricultural markets. Commodities tend to flow towards markets along lines of communication that have the advantage of relative cheapness. Changes in the technology and structure of transport networks which affected freight rates had the effect of changing the patterns of accessibility of urban centres. In France, the primary rail networks, constructed in the 1840s and 1850s followed the existing major axes of communication. Their configuration was basically unaltered. These remained the lines of force of spatial organization. Reduced costs led to an intensification of existing currents of trade, however, and to a greater concentration of commodity flows. As the geographer, R. A. J. Johnson, has stressed 'this is because each railway constitutes a traffic axis along which traffic (and therefore trade) can flow more cheaply in a linear path than it can be diffused over an economic landscape'. Within the railway network, the unit cost of transport tends to be lowest on the heavily utilized main lines, a factor which, when reflected in freight rates, induces further concentration of trade flows as areas in proximity to those main lines are offered access to markets at relatively low cost, and thus perceive a greater incentive to commercialize their production. Planned to converge on Paris, and to link major regional centres to the capital, the railways provided less effective links between the various regional centres and lower-order urban markets. They also increased the importance within the marketing system of the larger ports. Secondary axes of communication were less attractive in terms of cost and were subject to adverse competitive pressures. Thus, rail construction had the effect of increasing competition between regions so that, in spite of the basically unaltered configuration of the transport network, it did not follow that trade continued to take the same direction.

Some regions were particularly disadvantaged by delays in initial construction and the continued effect of difficult natural conditions. In upland areas, steep gradients limited the efficiency of rail transport. Modifications of transport networks thus had the basic effect of confirming existing disparities within the communications and within the marketing systems. Many markets of secondary importance experienced rapid decline because of the growing spatial concentration of commerce while, with some significant exceptions, existing regional centres underwent growth, based upon modifications in the relative accessibility of markets to potential clients – measured in terms of distance-time. Where the transport of goods was slow and expensive, then markets were scattered. Changes in the distance-time equation allowed a growing concentration of activity at central points which were more accessible and offered a wider range of services to users.

In the worst situation were those areas which, late in experiencing transport improvement, during the period of delay lost markets to producers provided more rapidly with improved communications. Thus, the most bitter complaints about improvements in the communications network and the effects this had on marketing came from areas which felt neglected. They were complaints about the lack of railways in a locality, and about the relatively high cost of reaching the nearest station, because of the distance and because the poor quality of local roads, from groups, such as the farmers at Royère (Creuse), who had traditionally produced cereals for sale in the Auvergne but found themselves in a less competitive situation in the 1860s as imports into the Auvergne by rail increased while they continued to depend on transport by ox-cart.[3] An enquiry conducted by the *Société national d'agriculture de France*, and reporting in 1873, observed that these complaints about poor access to markets were particularly frequent from the west and south-west.

To be distant from a railway station meant additional difficulties in marketing produce, which inevitably reduced the attractiveness of the market economy. As the municipal council of Fontaine-le Port (Seine-et-Marne) dramatically declared in 1872, it was to be 'without life'. It meant that outside merchants rarely, and only as a last resort, approached these areas either to buy or sell, because of the additional costs which resulted from activity there.

Most of the countryside, however, was gradually more fully integrated into the market economy. Before 1914, the railway was the only means by which this could be achieved. A parliamentary commission of enquiry in 1912 clearly recognized this in stressing the fact that 'regularity and low cost of transport are today the essential precondition for the economic prosperity of a region and even the development of its civilization. . . . Amongst the methods which might procure these advantages to a region, only the railways are susceptible of penetrating every part of the region.'

In addition to network configuration, the directions in which commodities flowed were also determined by deliberate commercial action by the individual railway companies, designed to increase the traffic using their lines. The 1840s were the formative period in the development of a commercial policy. Thus, the administration of the Paris-Orléans company noted in January 1847 that, whereas wheat was being sold at 30f47 per hectolitre in Paris, it cost 34f79 in Orléans. It was quickly realized that the correct tariff would promote a flow of wheat from Paris to Orléans and benefit the company and, in this year of dearth, consumers as well. It was clear that competitive reductions by companies in freight charges might cause repeated changes in market-structures. These were particularly likely to occur where railways faced waterway competition. The cost of marketing agricultural products was lower from areas in which effective competition existed than from those in which the railways enjoyed a decisive advantage. Thus, the farmers and millers of much of the Île-de-France complained about their inferior position compared with competitors in close proximity to the Seine valley. Complaints were also frequently made about the complexity of rail tariff structures and about their high level. The most frequent complaint, however, appears to have concerned the workings of differential tariffs, i.e. charges digressive with distance, and thus attacked the basic rationale of company freight rate policy. A differential tariff structure had positive advantages for the railway companies. In the first place, it increased their competitiveness and could be deliberately manipulated to do so where competition was most severe. Secondly, differential tariffs attracted increased traffic, and, given high initial construction costs and substantial fixed operating costs in respect of equipment and labour, it was important to use facilities to the maximum. Differential tariffs allowed the railway companies to relate their charges more closely to operating costs, for fixed costs made up a smaller proportion of the total cost the farther the goods were transported, so that the actual cost of moving a commodity, while continuing to increase with distance, did so at a decreasing rate.

For the rail user, differential tariffs had the effect of reducing the geographical disadvantages of distance or, conversely, of reducing the advantage of proximity to particular markets. Above all, they benefited Paris, the terminus for most lines, rather than intermediary points. They tended to favour departments distant from Paris – or, indeed, other major markets – at the expense of those in close proximity; and, in addition, the penetration inland of imports from the ports – and for this they were constantly attacked. In the 1860s, flour millers in the Nancy area were outraged at having to pay 32f per tonne to Paris while their competitors at Strasbourg enjoyed the benefit of differential tariffs and payed only 20f. The *Tribunal de commerce* at Meaux (Seine-et-Marne) pointed to the absurdity of a situation in which grain merchants supplying Paris might find it more profitable, because of differential tariffs, to purchase imported grain at Strasbourg rather than the produce of their own area. It was maintained by another observer that, in years when good harvests promoted price reduction on all European markets and when within France each market would otherwise have restricted its provisioning to local suppliers, differential tariffs had the effect of increasing long-distance shipments – particularly imports – which resulted in a further reduction of price levels.

The market for wine was also affected. The charges levied by the Paris-Lyon-Méditerranéen company, especially from the 1880s, tended to favour wine producers in the Midi at the expense of those of the Maçonnais and Bourgogne and of the Beaujolais. In the decade 1895–1905 it cost only 28f per tonne to transport wine from Sète in the Hérault to Paris compared with 32f for the much shorter distance from Belleville (Rhône). This increased the volume of flows of wine from the Midi towards northern France but at the expense of other regions.

Clearly railway freight rates were designed to encourage long-distance movement and, given the characteristics of railway cost structures, it made good sense for the companies to encourage the regular development of long-distance and large-scale use of railway facilities. This was a constant feature of their policy. In the absence of effective pressure-group representation of agricultural interests, and, in spite of state supervision, the railway companies were relatively free to determine the charges made for the transport of agricultural produce and to do so primarily in terms of their own commercial interests. Thus, the particular interests of each network in determining tariff policies influenced the directions taken by flows of goods and inevitably, given the existence of special and differential tariffs, some localities and some regions were favoured at the expense of others. Company freight-rate policies frequently modified established patterns of trade based upon geographical proximity. In this process, numerous small marketing centres lost their privileged intermediary roles, and, in regional terms, so did the centre-east – Bourgogne and Franche-Comté – as direct rail links between north and south replaced the more dilatory waterways. The official advisory body, the *Comité Consultatif des Chemins de Fer*, observed in 1896 that tariff proposals made jointly by the P.L.M. and the Midi companies showed 'no concern for economic conditions'. They were simply designed to improve the profitability of the two companies. Such criticism reflected the public-service conception of the railways' function held by many officials, and, from a more obviously self-interested stance, by transport users who felt that their interests were threatened by company policy. The companies themselves had constantly sought to minimize government intervention so that, although the Minister of Public Works had from 1849 enjoyed the right to reject company tariff proposals as against the public interest and certainly exercised an informal influence on their commercial policy, he did not have the right, save in exceptional circumstances, and most notably those of dearth, to impose changes in freight rates.

Market Structures

Exact measurement of internal trade flows is impossible in the absence of statistical sources to compare with those for external trade. Analysis of the impact of rail construction has therefore to be based primarily on the accumulation of information dealing with specific locations. According to the French economic historian, D. Renouard, the centuries-old process of national market integration was finally completed in 1863, when the average distance travelled per tonne of merchandise reached 197 kilometres, the same as in 1936.[4] Already by 1861–64, around 30 million hectolitres of cereals were being transported annually by the six major railway companies, amounting to almost one-third of total consumption. The wine trade and that in livestock were similarly affected. In 1863 the same companies transported over six million animals, including one-and-a-half million destined for the Paris market. Substantial imperfections remained in the market system, due in large part to the slowness with which some regions were provided with modern transport networks. Nevertheless, the years 1840–90 can be regarded as a key period of structural change in the economy in terms of improved access to markets and the creation of far more integrated national and international trading systems. Above all, it was the impact of more assured food supplies on price mechanisms, on price levels and on consumer behaviour which will have to be noted.

It should not be forgotten, however, that there was considerable continuity in the shape of commodity flows, particularly of products of relatively low value in relation to bulk, such as cereals, which continued to be cultivated in most localities. This needs to be stressed. Short-distance transport remained by far the most common, for the supply of local towns and neighbouring areas of deficit. High rural population densities and continued relative isolation resulted in the consumption of a substantial portion of the product of agriculture in the countryside itself, or in the mass of small market towns. Most farmers and merchants took the path of least resistance to traditional markets, sometimes responding to new opportunities but often only when forced to by developing competitive pressures. Large-scale producers were the exceptions in that, for them, sale to major wholesalers, who resold either for consumption in a regional metropolis or in more distant areas, might assume priority. Although the zone of provisioning of Paris and other major urban centres was significantly extended, a substantial proportion of their foodstuff needs continued to be met by the traditional areas of production – in the case of Paris by the farmers and millers of the Beauce, Brie, Vexin, Valois and the plains of Picardy. Members of the *Comité Consultatif des Chemins de Fer*, in discussions held in 1896, stressed that, even on the railways, movements occurred from a multiplicity of points, and primarily over short distances, with oats and barley for animal feed being transported further than cereals for human consumption because of the greater spatial variation in their price levels. Thus, movement of cereals from the Nord network on to the tracks of other companies represented only 6.5 per cent of the cereals transported by that company in 1905, and an abnormal 19.11 per cent in 1911, although the tendency was towards increasingly long-distance transport. Flour, of greater value in relation to weight, was transported over longer distances. In general, however, although substantial reductions in the cost of transport had occurred, and were of special significance in the aftermath of poor domestic harvests, the reduction was not large enough in relation to the price of cereals to promote vast inter-regional flows over long distances in 'normal' years.

It was not easy for contemporaries, nor is it for historians, to distinguish between the effects on agricultural prices of more efficient distribution, growing domestic production and rising levels of imports. The economist, D. Zolla's indices[5] relating wheat supply, population and price do not make a decision concerning the relative importance of these variables any easier but they do show how total supply was increasing more rapidly than population, and we know that per capita demand was stable.

Table 4.1 Indices relating wheat supply, population and price

Year	*Total supply*	*Population*	*Price*
1856–65	100	100	100
1866–75	104	101	114
1876–85	116	101.4	99.5
1886–95	122	103.6	83.5
1896–1905	119	104.5	74.2

Source: D. Zolla, *La Blé et les cereales* (Paris, 1909).

Clearly, from the late 1870s, decisive changes occurred in these relationships. Most notably, imports became more regular and substantially increased in quantity, a development causing considerable anxiety in farming circles.

Repeated shortages due to poor harvests in the 1840s and 1850s did much to encourage the liberalization of trade. Throughout the first half of the century, these had been followed by the temporary suspension of the sliding scale of customs dues, or at least by substantial reductions in the tariffs payable. During the 1850s, this had occurred following the harvests of 1853 and 1855. After the latter, the sliding scale had not been restored until May 1859, and was then again suspended by decree in August 1860. Furthermore, a clear warning was delivered by Rouher, the minister responsible, that the government was thinking of more permanent legislative changes. Thus, suspension of the sliding scale was followed by a law of 15 June 1861 which reduced the tax on entry to a nominal f.50 per quintal for cereals or flour carried in French ships or to 1f where the carrier was a foreign vessel. Tariffs on other agricultural products were also reduced. It was not until 1881, and in response to pressures we shall have to discuss, that protective tariffs were reintroduced.

The combined impact of free trade, the rapid transmission of market-place information by telegraph, cheap bulk transport by rail and falling oceanic freight rates was finally to break – in the advanced economies at least – the age-old link between climate, the state of the harvest and food prices which had largely determined the pre-industrial economic cycle. Whereas before the railways, the large-scale movement of cereals following a deficient harvest had imposed serious strains on transport facilities and substantially increased costs, with the railway, rapid and regular services were maintained, and charges, far from increasing, were reduced in recognition of the railways' public-service function.[6] It was during such years that improved communications networks made their most significant contributions in terms of assuring food supply and evening out price fluctuations. According to an official commission reporting in 1868, France had been saved from famine in the previous year only by the combination of free trade, the telegraph, steamers and the railway. Well might an economic journalist, contrasting the relative ease with which Paris could be provisioned in the railway age with the immense effort previously required, conclude that *La famine est désormais impossible en Europe!* The period of transition was not without its difficulties. Following the harvest failure of 1855, when it became necessary to import 10 million hectolitres of wheat, the uncertainties caused by the sliding scale of customs tariffs, together with the incompleteness of the rail and telegraph networks, meant that this still took eight months. Even so, the *procureur-général* at Angers was able to announce with delight that for the first time in history a year of shortage would pass without disorder. By 1861, when 12 million hectolitres were imported, provisioning was completed before the end of the year.

The railway was of decisive importance in easing the penetration of imports into the interior of the country. By land it made possible the transport of cereals from as far afield as Hungary, provided, of course, that the price differential between Hungarian and French markets was large enough to cover relatively high transport costs. In one of his regular situation reports, the *procureur-général* at Metz recorded his surprise at this development which he believed occurred for the first time in the last quarter of 1861. It was, however, the railway in combination with the reduction in maritime freight rates and the extension of cultivation and commercialization in the United States, Russia and India which provided the main competitive threats to French farmers.

Completion of the P.L.M.'s main lines in 1857 finally established a north-south rail link and made it possible for grain and flour merchants in the ports of Le Havre and Marseille to compete for the supply of vast hinterlands. Initially their zone of competition extended as far as the Paris market. As the volume of imports from North America channelled through the northern ports became relatively more important, however, they tended to push grain imported into Marseille further south. Merchants operating in the north benefited from lower shipping charges and easy access to Paris and other major markets. The freight charges levied by the railway companies were also of crucial importance. The transport of imports was an important source of revenue for the companies, particularly from the late 1860s. Differential tariff structures were devised to facilitate the penetration of imports inland. Charges were designed in part to encourage imports through ports which served the lines of particular companies. This policy was reinforced by a species of local protectionism extended to the products of areas with access to the lines of a given company where these faced competition from products in regions served by other companies. The objective was achieved by restricting the application of low long-distance tariffs and by agreements between companies to share traffic. These restrictions limited competitive pressures from outside the given 'zone of influence' of a particular company from both extra-regional and international producers. They were, of course, introduced only where the volume of imports was not sufficiently large for its transport to assume priority. The *Comité Consultatif des Chemins de Fer*, discussing such action in the early 1890s, accepted its legitimacy, but not without some of its members protesting, as did M. Cochery, in a session on 30 May 1894, that 'in a country like France, where the railway networks were established with substantial aid from the state, where the companies are in reality its delegates for a great public service, it is not admissible that the regions served by each network should be treated as separate countries'.

Imports obviously had their most significant effect following a poor domestic harvest. When harvests were good and domestic prices low, both inter-regional and international transport of cereals were restricted. Even so, the general improvement in supply led to heightened competition on major markets and to a levelling of prices, particularly in the Paris basin and in the Rhône-Saône corridor, in both of which regions internal flows of cereals and flour were supplemented by growing imports. The increase of supply in these two key areas had substantial direct and indirect effects on price levels and on commodity flows throughout the agricultural economy.

Direct effects were especially evident in the case of cereal producers in northern France, and in the centre, centre-west and east. The former faced competition from imports, particularly those brought along the Seine, and growing competition from domestic producers as production increased. The latter had by the 1880s almost been excluded from the Paris market owing to imports through Le Havre, while their zone of distribution towards the south was restricted by the direct and indirect consequences of imports through Marseille. *Indirect* effects were felt owing to the levelling of prices which tended to occur, 'initially between neighbouring zones', and then proceed rapidly as if by contagion from zone to zone. Although new flows did not directly compete on every market, they had the effect of excluding some producers from traditional markets, and of 'imprisoning' their surpluses locally, that is, of cutting off flows and leading to excessive local supply. This was the case, for example, in the Tarn in the late 1860s with the loss of markets in the Alps and in Hérault.

Before the establishment of the railway network, grains appear to have been moved mainly in a step-by-step manner, with the major flows taking place from north to centre, influencing supply and price levels in central France, and generating flows from there towards the south. After poor harvests, however, markets in the south had been invaded by imports through the ports. This had limited penetration from the centre and, indirectly, penetration of the centre from the north.[8] Already by 1866, that is to say, before the great period of import from North America, there were widespread complaints about the growing inability of grain brought to market at Dijon, for example, to proceed as far down-river as Lyon, much less to reach Marseille,[9] about the decline of sea-borne traffic moving from the north towards the south, and more generally about the effects of imports on price levels along the Mediterranean coast as well as close to the axes of penetration inland. Although Lyon seems to have remained the major point of contact between domestic produce moving south and imports, it appears that the competitive position of the former grew significantly weaker after the suspension of the sliding scale in October 1853. Exclusion of the cereals of the east from the Midi caused an increase of supply in Lyon even when foreign grains did not penetrate as far as the city. The responses to the 1866 agricultural enquiry from Ain, Côte-d'Or, Haute-Marne, Haute-Saône and the Vosges explained the decline in sales of their wheat in the Rhône-Saône corridor and in the Midi in terms of imports at Marseille; 'driving back' cereals towards the north, and also the establishment of new means of communication permitting areas closer to the Midi, but which had previously been isolated by inferior communications, to market their produce more easily towards the south. Some areas appear to have been caught in the middle. Thus, the Dauphiné Plains were increasingly excluded from markets further south by imports, and found it difficult to sell further north because of the growing surplus produced in the north-east.

If, after large domestic harvests, grain would again flow down the Rhône-Saône corridor towards Marseille and levels of import decline, this was only because of the very low level of internal prices, and the current was soon likely to be reversed. Contemporaries were particularly concerned about the greater instability of the market these various flows revealed. Producers in the north, centre and south no longer had guaranteed markets. Improved communications and free trade ended that. Farmers depended on merchants' assessments of the potential profitability of transactions conducted over increasingly wide areas. Merchants and millers bought grain with the object of maximizing profits, and clearly the interests of agriculture and those of commerce did not always coincide. The origin of cereals was not of crucial importance to commerce.

To some extent, the growth of competition was compensated for by the increased flexibility allowed by more rapid diffusion of information and improved transport. M. Tisserand, *rapporteur* for 1866 from Alsace, claimed all commodities 'to have acquired more fluid properties'. The declining significance of physical market-places and the growing importance of direct purchase on the basis of samples further contributed to this fluidity, so that produce might be moved in directions which 'varied according to the circumstances and sometimes very briskly from one year to another'. Despite continued market imperfection and the inertia of actors in the market-place – not altogether surprising in a period of rapid change – it was still easier to transfer activity from one market to another.

In these ways, the characteristics of markets were substantially transformed so as to provide new opportunities, but in a more competitive situation. Ambitions might be developed but, additionally, insecurity was increased. Thus, in 1866, the *comice agricole* at Craon (Mayenne), while welcoming the the opportunities created by improved access to Paris and to British markets, seemed more concerned about the appearance of livestock from the Rhineland on the Paris market, and especially the possibility that *one* day the Americans would export vast numbers of animals as well as wheat. This threatened 'the ruin of our agriculture'. Uncertainty, the feeling that everything was in flux, that traditional assumptions were no longer valid, inevitably caused anxiety. This nervousness was evident even in areas enjoying a unique prosperity because of improved communications. It was because, as the economist Bineau warned as early as 1843, 'the railways are changing too quickly the conditions of well-being, of prosperity, of existence, of those areas they cross'. Free trade intensified the problem of adaptation.

The acceleration of change was difficult to comprehend and to adapt to. Thus, the response to new market opportunities by farmers and by merchants was often muted. There was nothing inevitably positive about it.

A *'Crise d'adaptation'*

The second half of the nineteenth century was a time when most French rural communities experienced a major *crise d'adaptation*. New opportunities co-existed with the intensification of competition. External forces impinged upon rural societies to a far greater extent than ever before. They experienced a loss of autonomy. The price of locally produced commodities was henceforth dominated by prices on external, and even transatlantic, markets. The reduction in transport costs substantially modified the conditions for economic activity. Producers and merchants were to operate within a new spatial context, one subject to frequent change. The initial effect of improved communications, during the early years of network development into the 1860s, seems to have been to stimulate the trade in agricultural produce rather than to intensify competitive pressures, particularly as the first networks were constructed to provide links between major cities undergoing rapid population growth. As new commercial currents were organized, however, and as the density of the rail network increased and more localities were linked to it, so some regions experienced the loss of the formerly privileged position ensured by geographical proximity to major markets or water-borne access to them. This was clearly evident in some of the centres of supply to Paris – in Calvados, for example, which lost its pre-eminence in the provision of meat throughout the Paris basin. The social consequences were especially widespread in regions such as this, previously oriented towards production for the market and thus susceptible to changes in the structure of that market. The south and east of the Paris basin experienced a concurrent decline in the price of its two basic products – cereals and wool. In southern and central France, cereal producers with lower yields, but whose prices had previously been above the national average, were increasingly afraid of being driven out of production. In the 1860s, imported cereals appeared regularly in the important market at Toulouse as well as competing with its merchants elsewhere. Improved communications and free trade enhanced the mobility of cereals produced at home and abroad, at the same time as overall production was rapidly increasing. At least until the late 1870s however, rather than effecting a reduction in basic price levels, the primary contribution of the railway was to effect an evening-out in price fluctuations, thus reducing average prices, measured over quinquenniums or decades, but not 'normal' prices, determined essentially by costs of production. This was the change repeatedly noted by the authorities during the 1850s in relation to the disappearance of traditional subsistence crises.

The effects of improved communications on price levels were more substantial in the case of produce which had traditionally cost more to market – as a proportion of final selling price – than had cereals; in other words, *vins ordinaires*, meat, dairy products, fruit and vegetables. In these cases, prices initially tended to increase in response to a high elasticity of demand. In the past, for wine, as for wheat, prices had moved in inverse relation to quantities produced, particularly in the markets for *vins ordinaires* where variations in quality had little significance, and markets were geographicaly circumscribed by the cost of transport. With access to wider markets and growing demand, the mechanism was altered. Prices tended to be set in a national market and were, until the development of overproduction in the 1890s, stabilized at a relatively high level. Cycles of the type *shortage = high price* continued, but price rises were less marked because of more effective market integration; cycles of the type *abundance = low price* tended, temporarily, to disappear because of access to wider markets and rising per capita demand. This was initially a situation of great potential for profit, provided, of course, that the farmer produced a wine that was considered drinkable. Already in the 1860s, however, there was a growing awareness of the more competitive character of the market as production in the Midi increased.

In the longer term, and from as early as the late 1870s, price levels were to be depressed as supply began to exceed demand because of increasing domestic production and because of imports from Spain and Italy of wines that were often adulterated by the addition of alcohol to fortify them and of artificial colourants to improve their appearance. Subsequently, imports from Algeria intensified the problem. Again, penetration was facilitated by differential rail tariffs. The increase in imports during the phylloxera crisis limited the rise in prices which would otherwise have compensated producers for reductions in levels of production, and represented a significant obstacle to the process of capital accumulation which was a necessary prelude to the re-establishment of the French vineyards. Nevertheless, reconstitution of the vineyards with high-yield plants led to domestic overproduction. This was a clear case of the extension of markets and increased demand encouraging the entry of new producers and eventually leading to overproduction.

Previously transport difficulties had permitted the survival of vineyards on the climatic margins of production in northern France. These, and especially the vineyards of the Île-de-France, which had supplied Paris, declined rapidly once wine from Bourgogne, the Bordelais and Midi could be transported cheaply over long distances. Elsewhere, too, local wines for popular consumption were unable to compete with southern vineyards in either quality or priçe – in parts of Puy-de-Dôme, Corrèze, Cher, Moselle, or Alsace, for example. This often came as a considerable shock. The *comice agricole* at Semur (Côte-d'or) complained that sales 'of our wines of mediocre quality have greatly diminished since the establishment of the railway from Marseille to Lyon. The wines of the Midi offer us a disastrous competition.' The solution was either to improve the quality of the wine or to pull up the vines. In the centre and further south, in such departments as Aveyron, wine producers not only lost traditional markets, but even found that their local consumers increasingly preferred the wines of such areas as Hérault, Aude and Tarn with their higher alcoholic content. The wines of the Midi had the advantages of higher quality, lower costs of production and, in addition, the benefit of differential rail tariffs, all of which contributed to rapidly extending their markets.

The Demand for Protection

In these various ways continuity and change coexisted in the markets for agricultural produce. The efficiency of the marketing system appears to have increased, but within limits set by the high costs of operation of a declining though still large number of producers and intermediaries. The response of farmers and merchants to the accelerating process of change in market structures was gradual. Their awareness of change and of the possibilities for increasing productivity by means of technical innovation, or altering their product-mix, varied. Farmers were essentially 'price takers' rather tham 'price makers'. They were unorganized and in a relatively weak position within the market system. The improvement in communications and changes in commercial practices occurred at a speed that shocked and concerned most contemporaries.

The impact of changes in market structures was most strongly brought home to farmers by the clear tendency from the 1860s for agricultural prices to stagnate and the complex effects this had on incomes. The period between 1874 and 1895, that of the 'great depression', was to be particularly difficult because of the decline in the prices of most farm products. The crisis was especially marked for cereals, wine and wool producers, but affected, even if less severely, livestock and dairy farmers. Least affected were those farmers still producing primarily for subsistence, but even they needed to earn some cash.

The mass of the rural population had tolerated, rather than welcomed, the commercial treaties of the Second Empire. Subsequently, as productivity increased and farmers found themselves with growing surpluses to sell on increasingly competitive markets, discontent became more intense.

A protectionist campaign, backed by falling prices, was politically attractive and caused the authorities much concern. Already in 1859 it was claimed that opponents of the regime in the north were using farmers' anxiety to develop 'the spirit of opposition'. The regime was to be repeatedly accused of sacrificing rural interests to those of the towns because it favoured urban consumers with cheap bread, and because its programme of urban renewal was financed by rural taxpayers. Not that the Imperial government was to enjoy the political capital these policies were partly designed to secure. The rising cost of living in the cities, which appeared to coincide with the establishment of free trade, made even the urban population protectionist.

The growing volume of complaints and their increasing influence on the farming population was to be a matter of grave concern for governments as the agrarian depression deepened, and eventually required concessions. The rural elite of landowners and large-scale tenant farmers increasingly favoured protection, not simply out of consideration for the profit margins of its agricultural enterprises, but because these increasingly appeared too low in comparison with other forms of investment. By 1884 the Prefect of Haute-Saône could report that 'there is almost unanimity in favour of protection'. He added a warning that 'in spite of the sincerely held republican sentiments of the majority of the population, it is not likely that they will vote for anyone, however devoted to the republic he might be, who declared himself to be a partisan of free trade'. According to his colleague in the Moselle, 'regardless of the merits of protection, the government urgently needs to act in order to prove to the rural population that it is not indifferent to their interests as its enemies claim'. Protection was increasingly favoured, not only because of the effects of competition from imports but because of the possibility of the extension of such competition at some unspecified time in the future – from American wines, for example. Such was the lack of confidence and sense of malaise in agricultural circles. The solution to every problem, the universal panacea, appeared to be protection rather than technical improvement. Even where substantial innovation did occur, as in Picardy, agriculturalists still complained about the absence of protection, perhaps because their already high profit margins would be further increased in a domestic market from which their only genuine competitors, i.e. foreign producers, were excluded. According to one of the most active propagandists for protection, Auguste Mimerel, what agriculture needed above all, if capital was to continue to be invested, was confidence, and this required protection. Rather than accepting the conditions created by changing market structures, agricultural interests contrasted existing price levels with expectations based on past experience. Their perceptions of the present and hopes for the future were shaped by these expectations and the idealized construction of the past upon which they were based. The reintroduction of tariff protection by other nations, as the world agricultural depression deepened, was another powerful argument advanced by French protectionists. From 1881, a series of concessions were made to them with the introduction of a protective tariff on animals and meat. The nominal tariff on cereals was increased in stages to 7f a hectolitre by 1894. These did not entirely prevent imports although they reduced the volume and ensured that prices were higher than they might otherwise have been. Thus, at the end of 1911, for example, a quintal of wheat, which was priced at 18f42 on the Chicago market, cost 18f34 at New York, 18f87 in Brussels and 18f93 in London with the differences explained almost solely by transport costs. On the Paris market the price quoted was 27f50.

One corollary of this growing protection was to increase the number of profitable transactions that might be conducted internally by means, in particular, of movement from north to south. This involved the covering of regional deficit by the supply of French grains for which basic costs of production and of transport were higher than those of imports. The Ministry of Public Works made continued efforts to influence the railway companies to ensure that movement of domestic produce was assisted by more favourable freight charges. The terms of the Conventions negotiated between the government and the railway companies in 1883 required the latter to suppress all freight charges which favoured imports, except – and the exception was significant – where they were in competition with waterway carriers. M. Tisserand, on behalf of the *Comité Consultatif*, further recommended to the Minister in July 1892 that rail tariffs should be designed to encourage movements of agricultural produce from areas of surplus to those of deficit *within* France, and that to achieve this objective it was essential that the cost of transport plus charges by commercial intermediaries should not exceed normal price differences between regions, nor should the margin for profit be so small as to be rapidly eliminated by minor price fluctuations. A rate of 2f20 to 2f50 per 100 kilometres was suggested. Negative measures were also taken. Thus, when the P.L.M. proposed in July 1892 to reduce its charges for the transport of cereals and flour from Marseille to Paris from 28f to 22f a tonne, this was rejected in January 1893 and again in 1895 by the Minister of Public Works on the advice of the *Comité Consultatif* on the grounds that 'it is in reality a penetration tariff' which would have increased the competitiveness of imports in central France as well as in the capital. The return to protection was thus accompanied by the deliberate manipulation of rail freight rates in favour of domestic interests.

Official policy towards the railway companies, together with the effects of tariff protection, forced them into major revisions of freight rates. The companies, frequently acting in collaboration, sought to make the most of new or reinvigorated internal trade flows particularly of cereals from the north into the Paris market and towards Languedoc, and encouraged such developments by the introduction of lower charges. An official report in October 1896 indicated that this policy was successful in stimulating growing sales in the Midi by flour mills as far north as the Cher, Indre, Indre-et-Loire and Vienne, but that transport costs still excluded those from such major farming regions, further north, as the Beauce. New freight rates introduced in 1901 by the Paris-Orléans seem finally to have solved this problem and restored a semblance of unity to the national market. Thus, government and railway companies gradually and effectively responded to the often repeated requests from French agricultural interests for freight rate structures which increased their competitiveness with imports. To some extent, at least, the commodity flows from the ports inland which had built up during the free-trade years were reversed as the competitive position of domestic producers was improved. Of course, these measures also reduced the effectiveness of international market mechanisms, the pressures on farmers to innovate and the rate of structural change in patterns of land use and in farm size.

Yet protection had come to be seen by many contemporaries as a social, as well as an economic, necessity at a time when the still-large farming community was afflicted by a generalized crisis and needed to be reassured. Politically, it was hardly possible to stand against the tide. It is an illusion to believe that politicians, in the French context, could have resisted and asserted the need for unrestrained international competition as the means of stimulating long-term structural change in agriculture. The return to protection might then be seen as a rational response to the problems of French agriculture as it then existed, a result, rather than a cause, of its economic weaknesses. In effect, the transition to free trade, in combination with accelerating processes of international market integration brought about by improved communications, had been too rapid.

Conclusion

In spite of the policy reversal represented by the return to protectionism, market structures had been permanently transformed by the communications revolution and the greater integration of national and international markets as well as by the slow growth of regional specialization as farmers adapted to new opportunities and pressures. Within limits, the improvement of communications allowed the more efficient working of the market mechanisms beloved of liberal economists, enlarging markets, levelling out prices, encouraging specialization and innovation. As the *procureur-général* at Grenoble observed in 1866 –

> Until the last few years, in every province in France, one might almost say in every commune, the populations were constrained by necessity to cultivate all the essential foodstuffs. This was easy to understand, it was for them a question of survival, since transport difficulties would not have allowed them to find in other parts of France the foodstuffs they lacked. Even in this century we have seen departments suffer from shortage at the same moment as others complain about abundance. These facts are impossible today. Thanks to the railways and the means of communication of all kinds which criss-cross the territory of the Empire, not only has transport become easy but its cost has notably diminished. As a result a complete change in the habits of farmers must occur. Every part of France will . . . find it in its interest to consecrate itself almost exclusively to the production of whatever its climate and tools make most economic.

Previously, in most regions outside the north-east, specialization had been limited, not only by the widespread need to secure basic foodstuffs, but by poor access to markets and the absence of incentives. The farmers' first priority had been to achieve household self-sufficiency regardless of the prices for subsistence crops, and only secondarily to market any surplus. On most farms, cash crops were cultivated only if, within the overall agricultural system, they were compatible with subsistence crops. The opening up of the countryside by improved communications led to the more rapid development of specialization. It was no longer necessary to produce basic necessities everywhere. Regular supply could be assured. Therefore, it became possible to adapt agriculture more efficiently to the environment, and to market pressures. The process was gradual, but more rapid than ever before. The interrelated development of markets and agriculture had fundamental effects on both.[10]

New opportunities were an incentive to change, and competitive pressures forced change. The decline in the relative importance of transport costs increased the significance of other factors. The cost of production, rather than the cost of access to market, became the primary determinant of areal specialization. Even before the rail network had been completed, imports of wheat through Marseille were reducing cereals cultivation in departments like Aude and Hérault. The latter obtained about one-fifth to one-sixth of its needs from outside its boundaries in 1850 but, by 1862, this had risen to three-quarters, and Hérault was well on the way to becoming a region of vine monoculture. Improved communications and free trade had accelerated previous development. In this case, competitive pressures were compensated for by new opportunities. Here, and in other southern departments like the Gard, as one contemporary observer pointed out, 'the interest of agriculture is then clear; it must purchase elsewhere all the cereals needed', and cultivate crops better suited to natural conditions. Gradually, as in the Var and

Alpes-Maritimes, 'easier and more frequent relationships between farmers and the main centres of consumption ... help the former to understand which are the products whose sale would give them the most certain and advantageous benefits.' Specialization and increased production stimulated commercialization, and increased the integration of farmers into extraregional markets to buy and to sell.

The provision of modern communications did not necessarily, nor inevitably, produce a positive commercial or technical response from farmers. Everywhere it took time to organize commercial links, and, in the case of wine, to overcome consumer resistance based on an initial preference for local vintages. It took time to determine whether competitive pressures or new opportunities were temporary or permanent, and to accept the risks implicit in a long cycle of production – especially in the case of vines or fruit trees. The faculty for adaptation by agriculture in response to a growing market is less rapid than that of industry. Supply was less elastic. Regional imbalances in development survived. The most backward areas were characterized by inferior transport systems, but this was only one of the reasons for their continued backwardness, and not the most important if one bears in mind differences in natural endowment. Thus, the opening up of the Lozère and Haute-Ardèche or of the southern Alps was not favourable to the agriculture of areas which could not find a specialized product to exploit and whose traditional products faced growing competition.

To understand the limits to change in commercial practices and trade flows we need to understand the existing structure of agriculture and the willingness and ability of farmers to respond to new opportunities. It needs to be stressed that, with the exceptions of the Paris basin where specialization in cereals production increased, the Mediterranean wine-producing areas, and to a lesser degree part of the centre-west which increasingly specialized in livestock and dairy produce, there were relatively few instances of regional specialization. Substantial change occurred almost everywhere but it generally took the form of the minimum necessary to preserve the equilibrium of existing agricultural systems as competitive pressures built up, as migration reduced labour supply, and as the need to increase productivity increased the size of surpluses for sale and required a continuing increase in investment and in earning capacity. But, for most farmers, this participation in the market remained partial. Only a minority of efficient cultivators were *primarily* oriented towards the market and were in a position to compete effectively in an international market. For the others, price depression and tariff protection reduced the incentives to sell on the market, on the one hand, and the pressures to innovate on the other. The potential for change was reduced and its progress notably slowed.

This somewhat negative judgement, however, should not be allowed to conceal the fact that a massive change had occurred in the second half of the nineteenth century. It is easy to assess economic achievement in a particular country on the basis of comparison with apparently more advanced neighbours or with its own later achievements, but this is to ignore the particular structures and problems, the conditioning influence of period and place. However limited the consequences, it was during these years that the development of a modern communications network facilitated the creation of a market system which, if still in many respects imperfect, brought the traditional subsistence crises to an end. It was then that improved access to markets, together with the growth in competition, stimulated increased commercialization, improvements in productivity in sectors of the farming community, and that town and country came to be, for better or worse, far more closely united than ever before. It marked the belated end of the *ancien régime* in the countryside.

I would like to express my gratitude to Dr Colin Heywood of the University of Nottingham for his very helpful comments on an earlier draft of this paper.

Notes

1 B. Lepetit, *Chemins de terre et voies d'eau. Réseaux de transports et organization de l'espace en France, 1740–1840* (1984, Paris), p. 100. *See* also R. Price, *The Modernization of Rural France. Communications Networks and Agricultural Market Structures in Nineteenth Century France* (1983), Ch. 2, 'Pre-rail Communication Networks'.

2 1977. p. 89.

3 For further development of these points, together with detailed references, the reader should refer to Price, *Modernization.* Ch. 7, 'The Transport Revolution: railways, roads, waterways'.

4 D. Renouard, *Les Transports de marchandises par fer, route et eau depuis 1850* (1960, Paris), pp. 42–44.

5 D. Zolla, *Le Blé et les céréales.* (1909, Paris), p. 239.

6 *See* Price, Ch. 6, 'An End to Dearth'.

7 For the following section *see* Price, Ch. 8, 'Modernizing Market Structures'.

8 *See* Price, Ch. 3, 'Agricultural Market Structures before the Coming of the Railway'.

9 Ibid. Ch. 8.

10 *See* Price, Ch. 9, 'Agriculture in a Changing Market'.

References

Agulhon, M., Désert, G., Specklin, R., *Histoire de la France rurale*, vol. 3 (Paris, 1976).

Caron, F., *Histoire de l'exploitation d'un grand réseau. La compagnie du chemin de fer du Nord, 1846–1937* (Paris, 1973).

Clout, H. D., *Agriculture in France on the Eve of the Railway Age* (London, 1980).

Clout, H. D., *The Land of France* (London, 1983).

Cornu, P., *Une économie rurale dans la débâcle. Cévenne vivarais 1852–1892* (1993).

Demonet, M., *Tableau de l'agriculture française au milieu du 19e siècle* (Paris, 1990).

Farcy, J. -C., *Les Paysans beaucerons au 19e siècle*, 2 vols (Chartres, 1989).

Lepetit, B., *Chemin de fer et voies d'eau. Réseaux de transports et organisation de l'espace en France, 1740–1840* (Paris, 1984).

Margairaz, D., *Foires et marchés dans la France préindustrielle* (Paris, 1988).

Pfister, C., '*Fluctuations climatiques et price céréaliers en Europe du XVIe au XXe siècles*' *Annales E.S.C.*, 43 (Paris, 1988).

Price, R., *The Modernization of Rural France. Communications Networks and Agricultural Market Structures in Nineteenth Century France* (New York, 1983).

Price, R., *A Social History of Nineteenth Century France* (London, 1987).

5

Agriculture and Industrialization in France, 1870–1914

Colin Heywood

'A plague on agriculture and on agriculturalists! And especially on peasants: that swarm of small and medium farmers clinging to the land, to their feeble heritage, barely involved in the market, short of cash, bogged down in their routines, incapable of buying industrial goods on any scale, being preoccupied with getting more food, the poor people, and better food.' That is how Jean Bouvier caricatures a common attitude among economic historians to modern French agriculture.[1] In survey after survey, the peasantry is, indeed, depicted as an obstacle to overall development, contributing in no small measure to the alleged 'lag' in French industrialization. This negative view is particularly evident when the period 1870–1914 is considered. Take the influential article by Maurice Lévy-Leboyer on the deceleration of the French economy during the second half of the nineteenth century. It suggests that, from the 1880s, the economy divided under the pressure of increased competition. The urban sector invested and adapted, thereby benefiting from the challenge. But the rural sector ossified, thereby putting a brake on industrialization. All this is entirely plausible. Yet the whole argument can be stood on its head: an archaic agricultural sector may have held back industry, but it is equally logical to argue that slow industrialization held back agriculture. Vernon Ruttan, for example, asserts that the sluggish rate of growth in agriculture from the 1870s can, for the most part, be explained by the limited opportunities open to it in the French economy. Specifically, he points to the weakness of demand for agricultural products, the restricted outlets for productive employment in the urban-industrial sector and the reluctance to invest in an institutional infrastructure that could support agriculture.[2] Once again, though, there is the assumption of slow, retarded economic development in modern France.

This essay will contend that, to proceed further in the debate, historians need to take into account recent revisionist work on French economic growth. That is to say, the strengths as well as the weaknesses of the French economy should be considered, bearing in mind that, over the long term, the forces of dynamism overcame those of inertia. It will, therefore, present French agriculture between 1870 and 1914 in the context of a moderately successful case of economic development. It will also stress the interaction between the agricultural and the non-agricultural sectors, rather than have one dragging down the other. The work is divided into three parts. The first looks at the performance of the French economy, assembling the major indicators for our period. The second highlights two major influences on that performance: small farms and protectionism. The third concentrates on the linkages between agriculture and industry in France, a perspective which is often ignored or treated cursorily in the historical literature.

The Performance of the French Economy

How well did the French economy, in general, and the agricultural sector in particular, perform during the period 1870–1914? There is no shortage of data on which to base an answer to these questions. We should be clear at the outset, however, that the calculations made by quantitative historians must of necessity rest on the notoriously unreliable foundations of nineteenth-century official statistics. An element of 'spurious accuracy' must be admitted in the figures that follow, and so they should be considered at best a general indicator of economic performance. A useful starting point is provided by the annual average rate of growth of the physical product in France, calculated by Jean Marczewski as follows.[3]

Table 5.1 The annual average rate of growth of physical product.

	Gross agricultural product	*Gross industrial product*	*Gross physical product*
1865–74 to 1875–84	−0.3	1.6	0.78
1875–84 to 1885–94	0.0	1.5	0.91
1885–94 to 1895–1904	0.8	2.0	1.52
1895–1904 to 1905–13	1.0	2.0	1.64

Source: J. Marczewski, *Le produit physique de l'économie française de 1789 à 1913 comparaison avec la Grande Bretagne*, Cahiers de l'ISEA (*Histoire quantitative de l'économie française*), (Paris, 1965) Tables 33 and 35, pp. XCI and XCIII.

The last quarter of the nineteenth century stands out as a disastrous period for agriculture, with its negligible contribution to economic growth. These were, of course, the years often referred to as the Great Depression, when farmers in all parts of Europe had to come to terms with overseas competition on a scale hitherto unknown. French agriculture was particularly hard hit. As well as the influx of cheap cereals and livestock products from overseas, it had to contend with the phylloxera epidemic in vine-growing areas, a long-running crisis in the silk-producing regions of the south, and the disappearance of madder from the Vaucluse and surrounding departments in the face of competition from artificial dyestuffs. All of the local studies available confirm this impression of a general crisis on the land, particularly during the 1880s. In the Rhône department, for example, phylloxera took a heavy toll on the vines, affecting two-thirds of all plants at its peak in 1886. In the Soissonnais, big commercial farmers suffered most because of their dependence on the wheat market. Similarly, in the Pas-de-Calais the spread of market relations rendered farmers vulnerable to price declines, leading to stagnant output, declining profits and numerous bankruptcies.[4]

Industrial growth also slowed down at this period, but not to the same extent, and there was a more vigorous revival in this sector during the early twentieth century. Such a performance was insufficient to give the French economy as a whole much standing in the international context. A survey of Europe's Gross National Product by Paul Bairoch shows France being outdistanced by all her major rivals except Italy. Annual average rates of growth during the period 1860 to 1910 were as follows: Germany 2.57 per cent; Belgium 2.04 per cent; United Kingdom 1.87 per cent; France 1.41 per cent; Italy 1.05 per cent; **Europe** 1.88 per cent. A number of other indicators of economic performance, such as steel output or the total number of cotton spindles, can be used to place France 'low in the industrial league'. On the basis of this type of evidence, Clive Trebilcock confidently asserts that 'by 1900 France was clearly the economic laggard among the powers'.[5]

It may come as a surprise at this point to hear the conclusions of revisionist historians. They reject out of hand any talk of retardation, stagnation or backwardness. Instead, Rondo Cameron and Charles Freedeman contend that the French economy 'performed very well in comparison with other industrializing nations', while William Sewell suggested that it was 'one of the earliest and most successful cases of sustained modern industrial and economic growth'. To understand how this line of argument can be sustained, we must return to the quantitative evidence. The pivot of all revisionist writing is the per capita growth rate of the French economy. Using this measure, France emerges among the front runners. A reworking of Bairoch's figures to show annual average rates of growth of GNP per head between 1860 and 1910 produces these results: Germany 1.39 per cent; Belgium 1.12 per cent; United Kingdom 0.97 per cent; France 1.25 per cent; Italy 0.39 per cent; **Europe** 0.96 per cent. Bairoch concludes that his findings destroy the 'persisting myth of France's very slow economic growth'. In his view:

> The probable reason for the creation of this myth is the combination of three elements: very slow demographic growth (the slowest in Europe and for any developed country) . . .; the early start of France's industrialization when compared to other big European continental countries; and finally the fact that while France's industrial growth has been a little slower than that of countries like Germany, her agricultural production has increased more rapidly.

Some of the early enthusiasm of the revisionists may now appear unwarranted. As Crafts points out, in 1910, France ranked only seventh among European countries in terms of level of per capita income. Other historians highlight the limited amount of structural change in the French economy before 1914: industrialization, understood in the particular sense of the application of science and technology to production, was hardly impressive. None the less, France does emerge as a country that had some success in carving out her own path to development.[6] Describing this path is a little difficult, because it was neither particularly slow nor particularly rapid. A suitably nuanced summary of the 'French model of industrialization' comes from Jean Bouvier, portraying it as regular, progressive, dualistic, and without any periods of sharp acceleration.[7]

Once this overall achievement is recognized, a reorientation of approach to the performance of French agriculture becomes necessary. First, a balance needs to be struck between its strong and its weak points. A recent survey shows that around 1870 France (or, at least, northern France), together with Denmark, Britain, the Netherlands and Belgium, formed part of 'a core of very intensive and highly productive agriculture on the borders of the North Sea'.[8] It also reveals that between 1870 and 1910 the agricultural sectors in France and Britain were conspicuously slow to increase production and productivity, possibly because of the acute problems faced by large, capitalist farmers at this period. In the short term, then, French agriculture proved vulnerable to the series of internal and external shocks it received during the 'Great Depression' and its aftermath. In the longer term, however, as Bairoch's figures indicate, the gains in agricultural productivity were similar to those registered in other European countries.[9]

A second type of reorientation concerns the use and abuse of international comparisons. The notion of a French 'lag' behind a British or German model becomes difficult to accept. Different countries can be seen to be taking different paths to development, and so a priori there is no reason to expect France to have, say, as many large tenant farmers as the English, or steelworks on the scale of the Germans. This point is crucial for an understanding of our next section.

Table 5.2 Annual rates of growth

	1830–80	*1880–1910*	*1910–80*
France	1.1	0.9	3.4
Germany	1.5	2.2	–
Belgium	1.1	1.4	–
Italy	0.2	0.8	3.4
Netherlands	0.4	1.5	–
United Kingdom	0.7	0.8	3.0
Europe	0.6	0.8	2.9

Source: P. Bairoch, '*Les trois révolutions agricoles du monde développé: rendements et productivité de 1800 à 1985*', *Annales ESC*, 44 (1989), pp. 317–53; *idem* '*Dix-huit décennies de développement agricole français dans une perspective internationale (1800–1980)*' *Économie Rurale*, 184–6 (1988), pp. 13–23.

Key Influences: Small Farms and Protective Tariffs

Any attempt to explain the performance of French agriculture during the late nineteenth and early twentieth centuries must take into account two influences that loomed large at this period: the high proportion of small peasant farms, and the growing resort to protectionism. Historians emphasizing the French 'lag' have almost invariably seized on them as key sources of weakness. Small farmers, it is often alleged, were poorly equipped to face the challenge of heightened international competition, being short of capital and of technical knowledge. Milward and Saul, for example, argue that the '3 million peasant farmers' produced by the French Revolution of 1789 had not had much effect on the economy before 1870 but, at the end of the century, 'the forces of change collided with an immense and real obstacle'. As for the tariff barriers, these have for a long time attracted a powerful critique from free traders. Michel Augé-Laribé accused 'protectionist charlatans' of slowing the pace of change on the land, and encouraging farmers to become complacent in their inferiority.[10] In our view, there is some substance to these two sets of charges. On the one hand, however, they need to be set in the context of the French economy as a whole, rather than pinned to agriculture alone, and on the other, they need to be confronted with contradictory viewpoints.

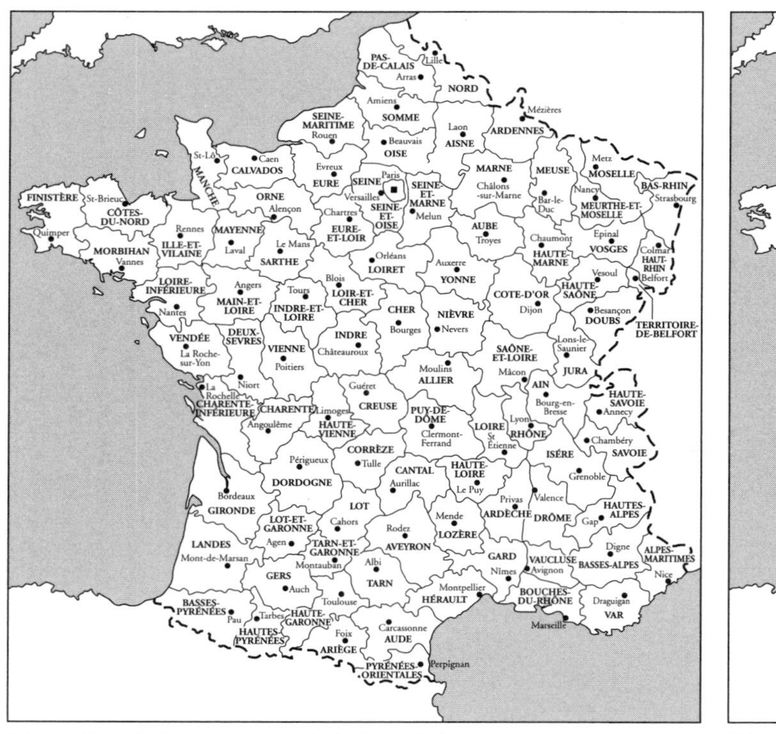

Map 1 **French departments and their capitals**
Source: **Eugen Weber,** *Peasants into Frenchmen. The Modernization of Rural France, 1871–1914,* (Stanford University Press, 1976).

Map 2 **French historical regions**
Source: **Eugen Weber,** *Peasants into Frenchmen. The Modernization of Rural France, 1871–1914,* (Stanford University Press, 1976).

A Republic of peasants

The role of small farms in the French system emerges clearly from the Agricultural Enquiry of 1882:

Table 5.3 The role of small farms in the French system

Size in hectares	*Number*	*Percentage*	*Area covered percentage (in thousand hectares)*	
1–5	1,865,878	53.2	5.6	11.5
5–10	769,152	21.9	5.8	11.9
10–20	431,335	12.3	6.5	13.4
20–30	198,041	5.7	5.0	10.3
30–40	97,828	2.8	3.4	7.0
40 and over	142,088	4.1	22.3	45.9
Total	3,504,322	100.0	48.6	100.0

Three-quarters of all farms were of less than 10 hectares in size: not enough land to support a family in most parts of France. It can always be pointed out that large farms of 40 hectares and over, although accounting for only 4 per cent of holdings, covered 46 per cent of the surface area. In other words, nearly half of the land was cultivated in big units of production, requiring the systematic use of wage labour. Here a comparison with Britain at the same period is instructive. In Britain, holdings of less than 20 hectares covered approximately one-fifth of the surface area; in France the equivalent was two-fifths. There were, of course, important regional variations. Closest to the British system were the Nord and the Paris basin, with their predominance of substantial tenant farmers. Late in the century, a similar structure began to appear in some of the wine-producing areas, notably in Languedoc. Another bloc, located in the west, south-west and parts of the centre, was characterized by large-scale property rented out in small units. This left approximately half of France with a widespread ownership of land and numerous small or medium-sized farms. The main areas for this independent peasantry were the east, much of the south, part of the west and the mountains.

If, in the late nineteenth century, France was something of a 'Republic of peasants', it could, with no less justification, be described as a 'Republic of artisans'. The 1906 census revealed a veritable swarm of industrial enterprises: over 2 million of them if the *travailleurs isolés* are included. Leaving aside the latter, one-third of all industrial enterprises employed less than ten workers, and the median was only 45 employees. Factories of a size comparable to other European nations were often to be found in mining, metallurgy and textiles. But small workshops featured prominently in clothes making, food processing, metal and woodworking. Taking the active population as a whole, in 1911 France a mere 46 per cent could be described as wage earners, compared to 90 per cent in Britain!

In these circumstances, a certain mystique became attached to the small producer. The drawbacks to the early phase of industrialization haunted some of the more sensitive observers in France, as they reflected on the relentless work regime of the factories or the squalor of slum housing. One reaction was to hark back to the supposed virtues of pre-industrial life, associated with the peasant and the master craftsman. This questioning of the benefits of industrialization was not confined to France, though it may be that her turbulent political history gave added impetus to the yearning for stability in certain circles. Pierre Baudin, for example, considered the moderate productivity of France a source of satisfaction. By avoiding the extremes of wealth and misery, he perceived the country to be settling into a comfortable mediocrity. Around the turn of the century, there were hopes of a *morcellisme industriel*, based on the use of internal combustion engines or electric motors in domestic workshops. Above all, there was an idealization of the small family farm, continuing the Jacobin tradition of the 1790s. In 1909 the Minister of Agriculture, Joseph Ruau, extolled the advantages of a land-owning peasantry, supposedly a guarantee of 'economic productivity, social equilibrium and political equality'.[11] From here it is a short step to portraying France as a *société bloqué* (stalemate society), resting on a tacit conspiracy between the middle classes and the peasantry of the Third Republic. Industrialization would be accepted, but only on the basis of a compromise between the old and the new society. Such a hypothesis, as advocated by Stanley Hoffmann, is temptingly symmetrical with the unobtrusive character of French industrialization.[12] Yet it needs some refinement, bearing in mind the findings of recent research.

The assumption that small-scale production was necessarily inefficient in the nineteenth and the twentieth centuries does not stand up to close examination, in France as elsewhere. A number of studies has shown that small firms could be more flexible in their production methods and more responsive to changes in demand than their larger rivals.[13] Certainly, the general trend in industry during our period was towards concentration, as many of the old handicraft trades, such as weaving and shoemaking, succumbed to competition from the factories. Some rapidly developing industries around the turn of the century were highly concentrated: aluminium and electro-metallurgy come to mind. But others were still dispersed in small workshops, notably the bicycle, aeroplane and car industries. In the latter case, France could boast the second largest output of motor cars in the world, after the United States, and this was an industry with no less than 155 different firms! There is even evidence from the early twentieth century of an inverse correlation between the average size of firm in a particular branch and the value added per worker. This is not so surprising when one bears in mind that comparative advantage in France lay at the quality end of the market, where the skills and good taste of her labour-force had maximum impact.[14]

But what of the small peasants? Could they too be highly productive? Ronald Hubscher asserts that they could, no less than the big tenant farmers. He cites an official enquiry of 1909, which pointed to better results on the small farms than on the large ones in a majority of departments, particularly those in southern France. Typically, in the Ain department the local *professeur d'agriculture* observed that small farms 'possess superior labour, without being quite as well equipped as the bigger farms; they make intelligent use of their machinery, and are better at manuring their soil, taking care of their livestock, and supervising the farm'. The key to success was *pluriactivité*: becoming involved in a variety of schemes, both industrial and agricultural, which could generate sufficient capital for the enterprise.[15] In the Dauphiné, for example, the peasantry resorted to seasonal migrations, becoming hawkers or harvesters; they lodged tourists; and they worked part time in a number of industries, such as silk, paper or metallurgy. Small farms (especially those over 5 hectares in size) were particularly suited to livestock rearing, the production of luxury wines, fruit growing and market gardening. As we have already noted, it was arguably the small farms on the Continent rather than the large-scale ones in Britain (and the Paris basin) that adapted most successfully to new circumstances after the 1870s. Once again, though, the regional dimension needs to be borne in mind. Where the market opportunities arose, and where soil and climatic conditions were favourable, peasant farmers proved adaptable to changing circumstances. Hence there were plenty of market gardeners around Paris and other big cities, flower growers in Provence, or even small vine growers in the Languedoc, who practised a highly intensive form of agriculture.[16] According to Augé-Laribé, French agriculture, no less than French industry, excelled at the quality end of the market: 'the best fruits, fresh, tasty vegetables, fine poultry, butter, a variety of cheeses, honey, thoroughbred horses, wines superior to all others in the world, and excellent brandies'.[17] But, in the more isolated regions of the west and south-west, the peasants would have to wait for decades until such opportunities arose, while in the Massif Central and the Alps there was a steady decline of agricultural activity. In the meantime, the peasants conformed more closely to the traditional stereotype, producing and consuming as independent family units.

Protectionism

If the easy assumption that small peasant farmers were an obstacle to economic development is open to challenge, so too is the well-rehearsed case against protectionism. The erection of tariff barriers around French agriculture in the late nineteenth century was a relatively protracted process. The 1860s and 1870s were decades of free trade in agricultural produce, following Napoleon III's 'economic *coup d'état*' with the Cobden-Chevalier Treaty. A revival of protectionist feeling in the early years of the Third Republic led to the first set of tariffs in 1881. These included substantial increases in the duties on livestock and livestock products, but there was no change in the regime for cereals and wine. Later in the 1880s tariffs on cereals began to be imposed, and those on livestock, meat and sugar were raised. A more important turning point was the Méline Tariff of 1892. Henceforth agriculture was well and truly protected, the minimum tariff running from 5 to 20 per cent according to the product. To round all this off, in 1897 the Government passed the *loi du cadenas*. This allowed it to modify, at its own discretion, duties on cereals, wine and meat. Agricultural raw materials, such as wool and flax, were exempt from duties under this system, as an encouragement to industry. In return, French farmers were rewarded with some of the stiffest tariffs on foodstuffs in Europe – though these were partially offset by declining transport costs in the late nineteenth century.

The agricultural interest was by no means alone in seeking protection from foreign competition. In fact, it was rather slow to jump on the bandwagon. The early running during the 1870s was made by industrialists, notably textile manufacturers from Normandy, the north and the east, combining with coal owners and iron and steel producers from the north, east and centre. Most farmers were still inclined to associate the prosperity of the third quarter of the nineteenth century with free trade. The onset of the *crise agricole*, highlighted by a disastrous harvest in 1879, caused a rethink. Even so, it was mainly the big landowners of the north and east who were affected at the outset. Their rents from large-scale farms in the 'wheat belt' around Paris spiralled down with the price of cereals during the late 1870s and the 1880s. Hence, they and their tenants had every incentive to try to stem the flood of cheap imports from the new world.

The small and medium farmers of the peasantry were not under the same pressure from abroad. The majority were largely insulated from the market by their self-sufficiency in the basic necessities of life. They might even have to buy food on the market in lean years, giving the same interest in low prices as other consumers. During the 1870s, therefore, the mass of the peasantry showed no enthusiasm for a return to protectionism. A few groups were even resolutely hostile. In so far as the smaller producers were involved in the market, they were likely to fare better than the big grain producers during the crisis. Peasant specialities, such as meat, dairy products and fruit, suffered less from foreign competition than cereals, and benefited from rising demand in the towns. Augé-Laribé cites *viticulteurs* producing fine wines and brandies, and Norman dairy farmers specializing in butter, as two groups eager to maintain their exports to Britain and elsewhere. It required a concerted effort by Pouyer-Quertier and his allies in industry, on the one hand, and the aristocratic landowners of the *Société des Agriculteurs de France*, on the other, to win over the peasantry to protectionism. The agricultural interest ended up by playing for safety in a seemingly hostile world. We are drawn back to the theme of a desire for stability among the élite in France. The Méline Tariff can be depicted as part of an agreement, half-conscious, half-unconscious, 'to protect all *positions acquises* and to manage and even limit economic development for the benefit and security of all established interests'.[18]

Does this mean that protectionism was on balance a negative influence on French economic development? Most economic historians would probably argue along these lines. There is usually some recognition that there were short-term benefits for the economy during the 1880s and 1890s. The tariffs gave a measure of support to agricultural incomes in the difficult circumstances of the *crise agricole*, and thereby encouraged a more buoyant market for industrial goods. But the longer-term influence is considered harmful. For Marczewski, the protectionist option prevented France from adapting her agricultural and industrial structures to the needs of the world economy, and thereby condemned her to modest growth right up to World War II. In the case of agriculture, the chief complaint against the tariffs is that they did nothing to help solve the problem of a technical lag behind other countries. This 'policy of least effort' is contrasted unfavourably with the decision in Britain and Denmark to expose farmers to the full force of overseas competition. While British farmers and agricultural labourers left the land in droves for jobs in the towns, French peasants were able to cling to their traditional way of life. Similarly, while Danish and German producers improved their technical knowledge and mobilized into co-operatives, their French counterparts languished in their ignorance and isolation.[19]

This all sounds very convincing, one must admit. But is it not the case that historians have been seduced by the charms of classical free-trade theory? Such an harmonious system, in which each country maximizes its income through peaceful trade with its neighbours, is hard to resist. None the less, it is worth asking what happens when attempts are made to translate theory into practice. More specifically, did the period of free trade in the 1860s and 1870s stimulate the development of the French economy? And did the reversion to protectionism in the 1880s hinder further progress? Bairoch concludes that experience did not bear out the hopes of liberal theorists. Indeed, the results were the complete opposite of what was expected. Economic growth, according to Bairoch, was slower during the free-trade period than during the protectionist periods before and after. Moreover, there was no surge in innovation and capital formation under the liberal regime. The only country to benefit from the experiment in free trade was Britain: the most developed of all the European economies in the mid-nineteenth century. Conversely, the 'less-developed' countries on the Continent experienced faster economic growth than Britain in the protectionist phase preceding World War I. Hence Bairoch concludes that the French government of the day was amply justified in returning to protectionism, particularly bearing in mind the predicament of agriculture.[20]

All in all, there is perhaps an air of unreality to much of the debate on protectionism. In the first place, political pressures made it difficult for a French government to avoid putting up tariff barriers in the late nineteenth century. Internally, protectionism can be depicted as part of a strategy to defend the capitalist system, in the face of threats from both left- and right-wing extremists. Externally, important competitors, such as Germany, Austria-Hungary, Switzerland and Italy, had already abandoned free trade by 1881, and, as the old French proverb says, 'When you are among wolves, you howl . . .' Only Britain stood out as the major exception but, by 1870, the British were exceptional again in the small size of their agricultural sector. Crafts calculates that only 20 per cent of Britain's labour-force was active in the primary sector, compared to a 'European Norm' of 40 per cent. France, with over half of her active population employed in agriculture, could hardly afford the brutal run-down of this sector experienced on the other side of the Channel.[21]

In the second place, there is the notoriously difficult problem of measuring the impact of tariffs. Several historians have shown that the French economy was far from asleep behind its barriers. New industries, such as motor cars, aluminium, rubber and electrochemicals, flourished from the 1890s. Between 1887–91 and 1909–13 exports of industrial goods doubled. Meanwhile, on the land, consumption of chemical fertilizers doubled during the 1890s, and there was some shift away from cereals towards livestock and market-garden produce. The environment created by tariffs may have helped these developments, but it is impossible to disentangle their influence from others at work in the economy. The work of Bairoch is certainly open to the criticism that he minimizes this type of specification problem. He stands accused of focusing too narrowly on a possibly arbitrary relationship between tariffs and the growth rate of the French economy. The safest conclusion must be to play down the role of tariffs, and place them in the context of a society going for growth on its own, essentially moderate, terms.[22]

Linkages Between Agriculture and Industry

Changing priorities

This brings us to the final, acid test for assessing the contribution of agriculture to the industrialization of a country: the workings of the linkages between sectors. Here we should keep in mind three types of interaction. First there are the products supplied by one sector to the others; second, the demand which people in one sector may have for the products of another; and third, the resources, such as labour and capital, that may be transferable between them. By the 1870s, France was emerging as a mature economy. This level of development dictated the priorities for agriculture. The age-old struggle to overcome the subsistence problem was on the point of being overcome, definitively. A steady, if unspectacular, rise in agricultural productivity since the 1820s, combined with the availability of cheap and abundant supplies from abroad, made it possible to fulfil a major role of the primary sector: feeding a growing urban labour-force. From the late nineteenth century onwards, meeting the demand for agricultural produce was not a daunting task in France. With the slowest-growing population in Europe, the country had fewer new mouths to feed than elsewhere, and there was the usual decline in the proportion of family budgets devoted to food as real incomes rose. This did not prevent French farmers from failing to meet a modest increase in the demand for their produce of 0.5 per cent a year during the 'Great Depression', as they struggled to come to terms with foreign competition.[23] By contrast, in the early twentieth century, safe behind their tariff barriers, their output was once again sufficient to satisfy demand. Food prices were higher than in Britain, but it is difficult to see this as an obstacle to industrialization because tariffs kept prices at a similar level in Germany. Ruttan concludes that over the period 1880 to 1930, the supposedly 'inefficient' French peasant provided the urban-industrial sector with more food per capita and at lower real prices.[24]

The increasing maturity of the French economy also rendered the flow of products in the reverse direction less important for overall development. The farm population could be expected to function as a market for industrial goods. By the turn of the century, however, the industrial and urban sector was beginning to acquire the capacity for autonomous growth. Once again, a distinction needs to be made between the years of depression in the late nineteenth century, and the period of prosperity during the early twentieth century. On the one hand, the agricultural crisis had a catastrophic effect on the purchasing power of the farm population: Marczewski calculates that its value declined consistently between 1870 and 1900. He went on to assert that the decline may have accounted for up to 60 or even 75 per cent of the reduction in the rate of industrial growth at this period, though recent research suggests that this exaggerates the role of agriculture in the economy. (Lévy-Leboyer and Bourguignon estimate that, by the late 1870s, agriculture accounted for only 37 per cent of the GNP). On the other hand, from the 1890s, tariffs helped support farm incomes, and the general revival of activity helped restore agricultural purchasing power. Industry, meanwhile, benefited from a resurgence of demand from overseas. On the eve of World War I, exports accounted for one-fifth of industrial output, compared to one-tenth around 1850.[25]

The 'releasing' of labour

What was particularly required from agriculture at this period was the 'release' of labour to the non-agricultural sector. The mobility of labour has proved a contentious issue in the French case. There is no disputing that the 'rural exodus' was a relatively muted affair compared to the British and German equivalents. It is also likely that many of the poorer peasants and farm labourers could have increased their incomes by moving from agricultural to industrial employment. For example, an enquiry of 1892 estimated that agricultural wages were on average only two-thirds of the level in industry.[26] Some historians have gone on to conclude that agriculture was retaining surplus labour, thereby obstructing the formation of a factory proletariat in the towns. Thus, Lévy-Leboyer speculates that wage increases in rural areas between 1840 and 1880 permitted many peasants to buy a plot of land. In the late nineteenth and early twentieth centuries, this class of small peasant farmers, which had only recently and with great difficulty acquired its holdings, was so attached to the soil that it refused the chance of a higher income in the towns. Other historians have put a very different slant on the evidence, concluding that, if peasants stayed on the land before World War I, it was because industry offered few outlets for them. The tendency in France was to assemble an élite of specialists in the workshops, rather than an army of unskilled labourers.[27]

A closer look at the literature suggests that it is less polarized than it first appears. On the supply side of the labour market, the influence of a 'passion for the land' in rural society is generally accepted. Emile Zola made it one of the themes of his novel *The Earth*: witness the Père Fouan as he struggles to come to terms with separation from the land:

> But something he did not say, although it came through in the emotion that he was trying to conceal, was his immense grief, hidden resentment and appalling heartache at giving up this land which he himself had so greedily

> cultivated, with a passion that can only be described as lust, and had then added to, with an odd patch of land here and there at the cost of the most squalid avarice. A single piece of land would represent months of a bread-and-cheese existence, spending whole winters without a fire and summers drenched in sweat, with no respite from his toil save a few swigs of water. He had adored his land like a woman who will kill you and for whom you will commit murder. No love for wife or children, nothing human: just the Earth.[28]

Even revisionist historians, who refuse to consider the peasantry as immobile, concede that a certain preference for life in the countryside helped channel the French economy on its particular path to development. O'Brien and Keyder conclude that the agrarian system of nineteenth-century France:

> can be perceived as a set of functional institutions which influenced the masses in France to restrain fertility and to remain on the land until the urban economy could provide them with real material comforts and a civilised environment to compensate for the loss of community and a way of life that so many working people seemed reluctant to abandon during the initial stages of industrialisation.[29]

When it comes to the demand side of the labour market, what the two sides have in common is a lack of evidence. Both confine themselves to bland assertions that industry either did, or did not, experience shortages of labour.[30] For a convincing argument, one would need material from various local labour markets. Charles Kindleberger gives a hint of what might be found. A brief survey of the regions indicates that Paris was always able to attract enough labour from the farms; some of the industrial areas of the north and east experienced a scarcity during the early twentieth century, which was covered by immigration from abroad; other industrial centres in the south managed to recruit adequately; and there were a few regions where labour shortages in industry coincided with redundant labour on the farms, as in the Vosges, Alsace and Brittany. Overall, he doubts whether these conditions inhibited total growth in France.[31]

Capital flows

Finally, agriculture might be expected to function as a source of capital for industry. The French state certainly used the tax system to transfer resources from agriculture into the construction of social overhead capital, particularly between the 1840s and the 1880s. Towards the end of the century, the agricultural crisis encouraged a flight of capital from the land into more liquid forms of wealth. In the Pas-de-Calais, for example, Hubscher reveals the wealthier members of rural society turning away from loans to local farmers, preferring various financial outlets in the towns. But here, as in other parts of France, it must be said that the rural 'bourgeoisie' showed little interest in industrial shares. Rattled, perhaps, by the bankruptcies and uncertainties on the stock market at this period, they went for '*placements de père de famille*' such as *rentes* or railway stock.[32] There was also a determined effort by the banks to tap the savings of the peasantry. In the Lyonnais, for example, Garrier describes how lawyers, the traditional moneylenders in the villages, handed over their networks of clients to the banks. Managers were recruited locally, and soon understood the need for discretion with their customers from the farms: a rear entrance leading on to a quiet backstreet was essential.[33] This flow of savings could hardly have made much of an impact on industrial investment, which generally relied on retained profits. But then most authorities agree that there was no real shortage of capital for industrial development in nineteenth-century France.[34]

More to the point, perhaps, was a shortage of capital in agriculture. There has always been a strand in the literature which bemoans any loss of labour and capital from the land, ignoring the linkages with other sectors. But, in this area, they do have a case. Augé-Laribé argues convincingly that the state could have done more to encourage agricultural credit, not to mention a general infrastructure of technical training and research stations. A start was made in all these areas before World War I, but it could not match the Danish or German achievement. Hence, French peasants were at a disadvantage as they tried to increase their productivity.[35]

Conclusion

Should one conclude that French agriculture provides an unsuitable model for others to follow? At first sight, this is the obvious line to take, and many observers have indeed pursued it vigorously. The *crise d'adaptation* of the late nineteenth century might well be taken as an ordeal by fire which cruelly exposed the weaknesses of peasant farmers in France. Arguably, they refused to be dragooned into the factories like the English farm labourers, preferring to eke out a miserable existence on the land. They failed to organize themselves into co-operatives like their Danish and German counterparts, holding instead to a rugged individualism. And, needless to say, they proved incapable of adopting the highly mechanized farming methods characteristic of North America, being confined in many cases to a traditional polyculture. In sum, they stand accused of sheltering complacently behind a formidable tariff barrier, content with their routines, their ignorance and their poverty. As 'proof', we have such indicators as low wheat yields by the standards of north-west Europe and low levels of geographical mobility. The agrarian system emerges as a key obstacle to industrialization in France.

But at this point one might retort: who really wants to end up working in a cotton mill or a steelworks? Who in their right mind dreams of living in a slum in Manchester or Essen? Facing *La Chienne du Monde* (poverty and hunger) in a village in Brittany or the Alps could be grim enough for anybody. Yet there were some advantages to the French path to industrialization, which lead one to suggest that it can be considered in its turn as a possible model. France did, after all, achieve a respectable economic growth rate during the nineteenth-century, in per capita terms at least. What the crisis of the 1880s revealed in the starkest of terms was the deep-seated desire in French society for security and stability. Change on the land and in the workshops was quite in order, but it would have to proceed according to certain ground rules. The French, in other words, had their own particular 'culture of production'. The industrial population had some inclination to specialize at the quality end of the market, incorporating a maximum of skilled labour. On this basis, it had some success in developing new branches and exporting its products. Hence, besides the big battalions of workers in the coal mines and the textile mills, there were the myriad small workshops of the Paris luxury trades or the metalworking industry. Similarly, the peasantry showed some preference for holding out on the land, rather than trying their luck in the towns.[36] This reflected a fierce attachment to the 'rural democracy' characteristic of the agrarian structure in the east and much of the south, as well as the limited employment opportunities in industry. The sentiment appears to have rubbed off on to some historians. Gilbert Garrier ends his book with the provocative thought, 'What would the world be without peasants?'.

Notes

1 '*Libres propos autour d'une démarche révisionniste*', in P. Fridenson and A. Straus (eds), *Le capitalisme français, 19e-20e siècle* (Paris, 1987), p. 16.

2 M. Lévy Leboyer, '*La décélération de l'économie française dans la seconde moitié du XIXe siècle*', *Revue d'histoire économique et sociale*, 49 (1971), pp. 485–507; V. Ruttan, 'Structural Retardation and the Modernization of French Agriculture: A Skeptical View', *Journal of Economic History*, 38 (1978), pp. 714–28.

3 J. Marczewski, *Le produit physique de l'économie française de 1789 à 1913 comparaison avec la Grande Bretagne*, Cahiers de l'ISEA (*Histoire quantitative de l'économie française*), (Paris, 1965), Tables 33 and 35, pp. XCI and XCIII.

4 P. Brunet, *Structure agraire et économie rurale des plateaux tertiaires entre la Seine et l'Oise* (Caen, 1960), p. 328; G. Garrier, *Paysans du Beaujolais et du*

Lyonnais, 1880–1970 (2 vols, Grenoble, 1973), I, p. 420; R. Hubscher, *L'agriculture et la société rurale dans le Pas-de-Calais du milieu du XIXe siècle à 1914* (Arras, 1979), pp. 581–94.

5 P. Bairoch, 'Europe's Gross National Product: 1800–1975', *Journal of European Economic History*, 5 (1976), Table 5, p. 83; *idem*; '*Niveaux de développement économique de 1810 à 1910*', *Annales ESC*, 20 (1965), pp. 1091–1117; C. Trebilcock, *The Industrialization of the Continental Powers, 1780–1914* (London, 1981), p. 112.

6 R. Cameron and C. Freedeman, 'French Economic Growth: A Radical Revision', *Social Science History*, 7 (1983), pp. 3–30 (p. 4); W. Sewell, *Work and Revolution in France* (Cambridge, 1980), p. 147; N. F. R. Crafts, 'Economic Growth in France and Britain, 1830–1910: A Review of the Evidence', *Journal of Economic History*, 44 (1984), pp. 49–67 (p. 52); R. R. Locke, 'French Industrialization: The Roehl Thesis Reconsidered', *Explorations in Economic History*, 18 (1981), pp. 415–33. For a recent survey of the debate, *see* C. Heywood, *The Development of the French Economy, 1750–1914* (London, 1992).

7 Bouvier, '*Libres propos*', p. 17.

8 J. L. Van Zanden, 'The First Green Revolution: The Growth of Production and Productivity in European Agriculture, 1870–1914', *Economic History Review*, 44 (1991), pp. 215–39. For a similar conclusion, *see also* P. O'Brien and L. Prados de la Escosura, 'Agricultural Productivity and European Industrialization, 1890–1980', *Economic History Review*, 45 (1992), pp. 514–36.

9 P. Bairoch, '*Les trois révolutions agricoles du monde développé: rendements et productivité de 1800 à 1985*', *Annales ESC*, 44 (1989), pp. 317–53; *idem*, '*Dix-huit décennies de développement agricole français dans une perspective internationale (1800–1980)*', *Economie Rurale*, 184–6 (1988), pp. 13–23.

10 A. Milward and S. B. Saul, *The Development of the Economies of Continental Europe, 1850–1914* (London, 1977), p. 116; M. Augé-Laribé, *La politique agricole de la France de 1880 à 1940* (Paris, 1950), p. 77.

11 F. Braudel and E. Labrousse (eds), *Histoire économique et sociale de la France*, vol. IV/1, *Panoramas de l'ère industrielle (années 1880–années 1970)* (Paris, 1979), pp. 242, 354.

12 S. Hoffmann, 'Paradoxes of the French Political Community', in Hoffmann *et al*, *France: Change and Tradition* (London, 1963), pp. 1–117.

13 J. Gaillard, '*La petite entreprise en France au XIXe et au XXe siècle*', in Commission internationale d'histoire des mouvements sociaux et des structures sociales, *Petite industrie et croissance industrielle dans le monde aux XIXe et XXe siècles* (2 vols, Paris, 1981), I, pp. 131–87; C. Sabel and J. Zeitlin, 'Historical Alternatives to Mass Production: Politics, Markets and

Technology in Nineteenth-Century Industrialization', *Past and Present*, 108 (1985), pp. 133–76.

14 P. O'Brien and C. Keyder, *Economic Growth in Britain and France, 1780–1814* (London, 1978), pp. 152–3; J.-C. Asselain, *Histoire économique de la France* (2 vols, Paris, 1984), I, pp. 197–201. For an opposing view, *see* M. Lévy-Leboyer and F. Bourguignon, *The French Economy in the Nineteenth Century* (Cambridge, 1985), pp. 57–72.

15 R. Hubscher, '*La petite exploitation en France: réproduction et competivité (fin XIXe siècle-début XXe siècle)*', *Annales ESC*, 40 (1985), pp. 3–34; Ministère de l'Agriculture, *La petite propriété rurale en France: enquêtes monographiques (1908–1909)* (Paris, 1909), p. 2; Y. Rinaudo, '*Un travail en plus: les paysans d'un métier à l'autre (vers 1830–vers 1950)*', *Annales ESC*, 42 (1987), pp. 283–302.

16 P. Barral, *Le département de l'Isère sous la Troisième République, 1870–1914* (Paris, 1962), pp. 113–15; N. Soubeyroux-Delefortrie, 'Changes in French Agriculture between 1862 and 1962', *Journal of European Economic History*, 9 (1980), pp. 351–400; Hubscher, *Pas-de-Calais*, II, pp. 479–81; Garrier, *Paysans du Beaujolais*, I, pp. 445–50; R. Pech, *Entreprise viticole et capitalisme en Languedoc-Roussillon du phylloxéra aux crises de mévente* (Toulouse, 1975), *passim*; M. Hau, '*La résistance des régions d'agriculture intensive aux crises de la fin du XIXe siècle*', *Economie rurale*, 184–6 (1988), pp. 31–41; Augé-Laribé, *La politique*, p. 263.

17 Augé-Laribé, *La politique*, p. 263.

18 *Ibid.*, p. 66; M. S. Smith, *Tariff Reform in France, 1860–1900* (Ithaca, 1980), p. 239. For a broad perspective, *see also* J. V. Nye, 'The Myth of Free Trade Britain and Fortress France: Tariffs and Trade in the Nineteenth Century', *Journal of Economic History*, 51 (1991), pp. 23–45.

19 Marczewski, *Le produit*, p. CXXII.

20 P. Bairoch, 'European Trade Policy, 1815–1914', in P. Mathias and S. Pollard (eds), *The Cambridge Economic History of Europe*, vol. VIII, *The Industrial Economies: The Development of Economic and Social Policies* (Cambridge, 1989), pp. 1–160.

21 Crafts, 'Economic Growth', pp. 51–6.

22 Asselain, *Histoire économique*, I, Ch. 4; *idem*, '*Croissance, dépression et recurrence du protectionnisme français*', in B. Lassudrie-Duchêne and J.-L. Reiffers (eds), *Le protectionnisme* (Paris, 1985), pp. 29–53; T. Kemp, 'Tariff Policy and French Economic Growth, 1815–1914', *Revue internationale d'histoire de la banque*, 12 (1976), pp. 147–55.

23 Calculation based on Marczewski, *Le produit*, Tables 31, 42 and 43; and Ruttan, 'Structural Retardation', Table 3.

24 *Ibid.*, p. 718.

25 Marczewski, *Le produit*, pp. LXIV-LXIX; Lévy-Leboyer and Bourguignon, *French Economy*, p. 53; Asselain, *Histoire économique*, I, pp. 181–3; *idem*, '*Croissance, dépression*', p. 48.
26 C. Kindleberger, *Economic Growth in France and Britain, 1851–1950* (Cambridge, Mass., 1964), pp. 225–34.
27 Lévy-Leboyer, '*La décélération*', p. 502; P. Hohenberg, 'Change in Rural France in the Period of Industrialization, 1830–1914', *Journal of Economic History*, 32 (1972), pp. 219–40. Cf. Ruttan, 'Structural Retardation', p. 720, and Y. Lequin, *Les ouvriers de la région lyonnaise, 1848–1914* (2 vols, Lyon, 1977), I, pp. 43–4.
28 Emile Zola, *The Earth* (Harmondsworth, 1980), p. 38.
29 O'Brien and Keyder, *Economic Growth*, p. 194.
30 Cf. T. Kemp, 'Structural Factors in the Retardation of French Economic Growth', *Kyklos*, 15 (1962), pp. 340–1, and A. Beltrain and P. Griset, *La croissance économique de la France, 1815–1914* (Paris, 1988), pp. 47–8.
31 Kindleberger, *France and Britain*, pp. 234–8.
32 Hubscher, *Pas-de-Calais*, pp. 845–7; A. Daumard, *Les fortunes françaises au XIXe siècle* (Paris, 1973), *passim*.
33 Garrier, *Paysans du Beaujolais*, I, pp. 501–12.
34 Kindleberger, *France and Britain*, p. 68; Braudel and Labrousse, *Histoire économique*, IV/1, pp. 37–40 and 265–72; Asselain, *Histoire économique*, I, pp. 209–10.
35 Augé-Laribé, *Politique agricole*, *passim*; Braudel and Labrousse, *Histoire économique*, IV/1, pp. 373–5; Ruttan, 'Structural Retardation', p. 719; Van Zanden, 'Green Revolution', p. 237.
36 For conflicting views on this point, *see* the debate between C. Nardinelli and G. Schmitt in the *Journal of European Economic History*, 17 (1988), pp. 427–34 and 20 (1991), pp. 91–115.

Further reading

Magraw, R., *France, 1815–1914* (London, 1983), Chs 6 and 9.
Tracy, M., *Government and Agriculture in Western Europe 1880–1988* (3rd ed., New York, 1989).
Lebovics, H., *The Alliance of Iron and Wheat in the Third French Republic, 1860–1914* (Baton Rouge, 1988).

6

Italy – The Eternal 'Late-comer'?

Paul Corner

More than 30 years ago, Alexander Gerschenkron published his *Economic Backwardness in Historical Perspective*, stirring up debate and encouraging widespread rethinking about the processes of industrialization.[1] In Italy the impact was immediate – for obvious reasons. The problems experienced by the newly unified Italy in her progress towards industrialization were very much those identified by the American scholar; Italy appeared to be almost the perfect 'late-comer' of Gerschenkron's analysis. The ensuing arguments and investigations – about levels of accumulation, the role of the State, the degree to which Italy had, or had not, realized any kind of substitution of factors – turned out in the main to be fairly inconclusive. It was hardly possible to identify anything that could be termed 'take-off' before the modest economic acceleration of the first decade of this century; it was difficult to assess the degree of capital accumulation in agriculture; and it was unconvincing to attribute many of the changes that did take place to the intervention of the State.

If anything, the impression that rested was that Italy had even managed to be an unsuccessful late-comer. And this was an impression confirmed by many of the standard works on the subject, which tended to highlight weaknesses, deficiencies, delays and shortcomings. With the exception of only one or two fairly short periods, it appeared difficult to identify consistent progress; stagnation, crisis and recession remained the dominant terms. Until, of course, the 'miracle' of the late 1950s, which justified its name precisely by the very unmiraculous nature of what had gone before.

Italian agriculture was assigned a particular place of honour in this tale of woe. With the exception, of course, of the highly successful capitalist 'high farming' of the Lombard plain and the reclaimed lands of Emilia, there seemed few points of cheer. In general terms, agriculture was considered to be backward, inefficient, unreformed and even – especially in the latifundia of the south – semi-feudal. For Emilio Sereni, one of the most distinguished of Marxist historians of the post-World-War-II period, much of Italian agriculture was 'a ball of lead tied to the foot of Italian capitalism'[2] while, for another, Pietro Grifone, the persistence of a backward agriculture constituted one of the 'original weaknesses' of the modern Italian economy.[3] Above all, perhaps, the dominance of peasant cultivation of one sort or another (small proprietors, leaseholders, sharecroppers) was considered especially damaging, limiting the internal market through reliance on subsistence, lowering levels of productivity and failing to provide that much-needed capital for industrialization. Small wonder, it seemed, that Gramsci's criticism of the Italian State began precisely from the 'passive revolution' of 1859–60 when the new State decisively turned its back on any kind of agrarian reform. The slow, rickety, and distorted progress of Italian industrialization appeared to have its roots here.

Peasants and Industrialization

There is, perhaps, little that is surprising in this analysis. Traditionally, peasants and industrialization have been thought to be poor bedfellows. The peasant mentality, usually depicted as closed and conservative, more disposed to subsistence farming than to consumption through spending, is considered to square badly with concepts of market, enterprise and risk. It is not surprising that it has become almost axiomatic that the first step in a process of successful industrialization should be the destruction of the peasantry. There are – unfortunately – enough relatively recent historical examples of this line of thought to demonstrate its hold on political leaders intent on rapid economic transformation.

It is all too obvious that such ideas derive from a precise historical example – that of Britain, conventionally considered to have seen both agricultural and industrial revolutions in the period between 1700 and 1830. The problem is that, over the years, the specific example has become a generalized model. As the First Industrial Nation, Britain has, not surprisingly, provided the benchmark for the study of industrialization in other countries. In a sense, therefore, almost all the literature on industrialization has become comparative; either implicitly or, more usually, explicitly, success or failure has been measured, and often explained, in terms of the degree to which other countries were able to follow the British model. This is best illustrated by the very terminology of the argument, in which Britain is the 'first-comer' to industrialization, while other European nations play the role of 'second-comers', 'late-comers', or 'late-joiners' on the path already trodden by Britain. Indeed, it has to be said that, all too often, the problem that historians have addressed seems to have been that of identifying Continental European variations from the British model to explain the defects of Germany, Russia, France or Italy. Through the use of the language of the race or competition – slowness, lateness, backwardness, retardation – a picture emerges in which, in the Olympics of industrialization, Britain won the gold, others were placed, and others were merely also-rans. The implicit assumption in this vision of things is, of course, that all were participants in the same race.

In all these analyses, one thing is always certain: favourite candidates for the 'backwardness' medal are France and Italy. France – *par excellence* the country of the peasant-proprietor – comes in for particular attention. As Robert Aldrich has put it recently, 'Historians have traditionally pictured the French economy down to 1945 as a still life. Demographic and economic growth are seen to occur only very slowly when at all; the persistence of artisanal industry, polycultural farms, and a fragile financial system are major themes; miserly landlords, hesitant entrepreneurs, careless politicians, or infertile couples are held responsible for the greyness of the pictures.'[4] Richard Roehl notes that the keywords in most approaches to the French economy in the nineteenth century are 'retardation' and 'stagnation', and quotes Sir John Clapham to the effect that 'France never went through an industrial revolution'.[5] Even less, of course, according to this interpretation, did France go through that agricultural revolution which would have removed its class of peasants. 'France' – wrote Alfred Cobban in 1959, in what was clearly intended to be a self-explanatory comment on poor economic progress – 'is after all a republic of peasants.'[6] Indeed, the explanation provided by almost all those who emphasize French backwardness is that of the continued existence of the *paysannerie* – the mass of small peasant proprietors who, it is argued, reinforced by the Revolution of 1789, succeeded in blocking agricultural change for more than a century, thus ensuring that French industry languished with neither capital nor market. Here differences from the British model are somehow taken to be *per se* evidence of less successful economic growth, and, more often than not, the blame is laid at the door of agrarian structure and agricultural performance.[7]

If France is often accused of backwardness, Italy is certainly not far behind. A great deal of the writing on Italian industrialization after 1860 reflects the same sentiments. The fact that governments of the new Italy did not effect any kind of agricultural reform but tended, if anything, to consolidate the positions of large rentier landowners and allegedly inefficient *latifondisti*, has frequently been seen as a cause of delay in Italian efforts to catch up with the industrialization of the rest of Europe. Lack of dynamism in agriculture – it is argued – was reflected in low levels of accumulation and correspondingly low levels of growth in industry. Governments were forced to rake in money through indirect taxation, which hit peasants hard, thus compressing consumption and ruining the internal market. Here again, agrarian structures are seen as decisive in promoting or impeding economic growth. When Gramsci identified the origins of many Italian problems in the purely 'passive revolution' of the *Risorgimento*, he was thinking primarily of the political consequences that ensued from the confirmation of existing social and economic structures, but it is enough to read the debate that developed in the 1960s around this interpretation to understand that the issues of accumulation, capital formation and agricultural reform during the first phase of Italian development have remained central to the arguments about Italian industrialization.[8]

With the exception of Rosario Romeo, who argued that agrarian reform would have slowed down Italy even further by employing in the primary sector funds essential for industry, most writers have accepted the view that an inefficient agriculture, dominated in many areas by absentee landlords and rentiers, represented a dead weight on the economy, with a high percentage of the population failing to produce more than required for its own survival. One suspects that, underlying such opinions, the British model (which, after all, became the Marxist model) was never far away; the implicit assumption was that industrial revolution should be based on agricultural revolution, and that to attempt the first without the second was to get off to a very bad start indeed.

As is obvious, in all these debates the peasantry has had a very bad press. Just as the speed with which Britain deployed its labour out of the agricultural sector is seen as a major factor in promoting industrialization, so the retention of a high percentage of the population in agriculture – particularly peasant agriculture – has been considered a principal cause of backwardness in many European countries. Thus, getting rid of peasants has become almost a condition of industrialization; some countries did, some did not, but those who did not could hardly expect good results.

Yet dissatisfaction with this approach has been increasingly evident over recent years. This has been stimulated by several considerations. First and foremost, perhaps, has been the growing conviction among economic historians that the 'classic' picture of the British Industrial Revolution is in need of fairly drastic revision.[9] It is now quite widely accepted that the term 'revolution' is misplaced in the case of both agriculture and industry if – as has often been the case – it is taken to suggest cataclysmic change. Rondo Cameron remarks that the term has led to the belief in the apparent need for sudden, discontinuous economic change as the necessary prerequisite of a modern industrialized economy, whereas 'In fact, gradual industrialisation has been the norm, explosive growth the exception'.[10] N. F. R. Crafts underlines the point: 'the acceleration in the trend rate of growth during the British industrial revolution was more modest and gradual than was widely believed in the heady days of the "take-off" literature'.[11] Such comments serve to reinforce the position of those who prefer to argue that British industrialization was more a special case than a model capable of strict imitation.

More important perhaps, the questioning of the paradigm invites us to reconsider the whole issue of 'backwardness'. As used in the historical debate, the term has often assumed an almost moral tone. Britain was first and therefore best (the 'winner' in the race); other nations were 'backward' or 'retarded' and, therefore, inferior ('losers' or competitors who used the political and economic equivalents of steroids to reach the tape). In this way, temporal differences become also differences of quality. As one writer has put it, speaking of the French case, 'It is often assumed – when not explicitly stated – that France is backward, retarded, behind the times, or old-fashioned, *and that this is bad*'.[12] Collective Italian breast beating over delays, failures and weaknesses seems to derive from the same line of reasoning.

The revisionist argument should not be mistaken for an attack on comparative history of industrialization as such, however. But it is an attempt to redefine the terms of comparison. The British model has too often been used to suggest that other roads to industrialization are somehow 'anomalous' or genetically flawed. This has two probable consequences. The first is that features, which are in fact important general features of European industrialization, are overlooked or considered without significance, precisely because already subsumed to the British model; the second is, as suggested above, that national and regional peculiarities – what is specific to a particular nation, to a particular region, or to a particular industry – are not valued for their own specific qualities, which may make them unique, but are treated as simple deviances from the norm. There is a strong case for arguing that, for an understanding of the process of industrialization, the study of relatively backward areas of Europe – areas which came late to industrialization and by a different road – are potentially as illuminating as are the studies of the major industrial regions.[13] This is a view echoed by Cameron when he writes that 'there was not one model for industrialization in the nineteenth century – the British – but several'. To ignore this, he continues 'distorts the historical record [and] conceals the distinctive varieties of industrialisation'.[14]

O'Brien and Keyder are among those who have gone furthest in the attempt to change the terms of the argument on industrialization in Europe. Widely criticized for certain aspects of its methodology, their book's basic premise – that there is no paradigm of industrialization against which all experiences should be judged – remains of great value. The authors are at pains to stress that experience is not necessarily always transferable; the French simply followed a *different* path to industrialization – a path more suited to, or dictated by, the particular French context. In a significant passage, they conclude that there is no justification for the view that 'there is one definable and optimal path to higher per capita incomes and still less [for] the implicit notion that this path can be identified with British industrialisation as it proceeded from 1780 to 1914';[15] 'France cannot be fitted into a typology of European industrialisation, and its development reminds us that there is more than one way of transition from an agricultural to an industrial economy and from rural to urban society. Nor is it at all obvious that the path of economic development taken by France from 1780 to 1914 was inferior to the vaunted British model.'[16]

One feature of this analysis that is particularly notable is that relating to the role of agriculture in the economic development of nineteenth-century France. As we have seen, critics of French performance have always pointed an accusing finger at the alleged inefficiencies of the agricultural sector, dominated by peasant landowners who clung tenaciously throughout the century to their small plots, farmed by underemployed family labour. Certainly, by British standards, a disproportionately high percentage of the French population was employed in agriculture for the whole period, and structural change in employment between sectors was very slow in coming. But while, in the British case, economic growth seems to have been in part dependent on the speed with which the relative agricultural population fell, thus opening the way to very high levels of labour productivity in the sector, in France the retention of labour in agriculture does not appear to have had the expected disastrous consequences for national per capita income. It may be, of course, that French manufacturing was held back by the slow rate of agrarian change and that equivalent per capita incomes were achieved despite, rather than because of, support received from the primary sector. But this has really still to be demonstrated. Colin Heywood writes that 'critics of the French peasantry have been too hasty. They have seen an inefficient allocation of resources without providing evidence of how they could usefully have been redeployed in the conditions of the nineteenth-century economy; they have prescribed innovations and investments without showing how market forces would justify them; and they have insisted on a refusal to take up opportunities, without proving that the opportunities existed.'[17] In other words, it is not at all self-evident that French industry – much less factory based than British – could have absorbed with benefit a much higher rate of exodus from the land.

What is significant is that developments in some way comparable with England (again, it is necessary to repeat, judged in terms of per capita income) appear to have been realized without any agricultural revolution (until late in the last century), without the appearance of large-scale capitalist agriculture, and with the persistence of a high proportion of the working population in rural occupations.

Conclusions of this kind evidently put in question the generalized model of the relationship between agricultural change and industrial growth. The many re-examinations of the French case suggest fairly conclusively that the orthodox model is far too schematic, and that it is possible to associate the persistence of certain apparently conservative social forms in agriculture, such as the peasantry, and a consistent level of economic growth.[18] This is not only because such agriculture can itself be profitable (the reader will not need reminding that some parts of French agriculture have traditionally produced very high-value products), but also because the kind of industry that was developing did not require extraordinarily high levels of capital investment. Again, in the French case, this would seem to correspond to the fact that the small, artisan workshop was more the norm than the exception; it produced relatively high-value goods, which required less capital for mechanization and which felt much less the constrictions imposed by the limited market for its products. Thus, the different industrial mix in France and Britain ensured – not that there was no net transfer of capital between sectors in France, which there must have been – but that the level of this transfer was very different. This in turn determined that agriculture had very different roles in the two countries in the process of industrialization.

Italian Agriculture: Traditional and Modern

Put together, the revisionist position on industrialization – already even described as the 'new orthodoxy' – and the fresh approach to the question of 'backwardness' provide a stimulating starting point for a new look at Italy. Of particular importance in the revisionist approach is the relaxation in periodization, suggesting that almost all processes of industrial growth have been slower, patchier and much more gradual than previously thought. Richard Tilly makes the point very well, writing about Germany, but in terms that might equally be applied to much of the literature on Italy: 'There is a need to relax (or forget) the assumption that modern economic growth takes the form of a "big spurt" discontinuity associated with rapid, dramatic industrial changes, a need to see modern economic growth as a more gradual, longer-run process extending well into the pre-industrial period'.[19] When related to the Italian case, this approach invites a radical change in perspective. The research agenda, determined in large part by Rostow and Gerschenkron, begins to look somewhat inappropriate. Considerations based on the longer view inevitably require the scholar to treat with caution the idea that everything ought to have been crammed into a few years and invite scepticism over the supposition that nation formation and industrialization were in some way necessarily parallel phenomena. The assumption that Italy should have proceeded along the lines of, and at the same pace as, Germany can also be abandoned. As a result, the question at issue is no longer that of the failure of big industry to 'take off' rapidly into sustained growth in the years following Unification; the search for the 'big spurt' becomes less central and less urgent; and possibly even the role of the State needs to be at least in part revised. And, as paradigms fade, stage theory withers and 'big spurts' disappear, other factors begin to assume a fresh significance. Here the French example indicates very strongly that the place of agriculture should once again be reassessed.

In conventional terms, it would seem difficult to argue for a revision of the role of Italian agriculture in industrialization. No-one contests the efficiency and profitability of the capitalist agriculture of the Po Valley and of other specialist regions, but these areas were clearly unable to compensate for the less-developed zones of the rest of the peninsula. Labour productivity undoubtedly remained low in many parts of the country. O'Brien and Crafts agree that one of the key, and incontestable, features of the British example was the speed with which labour was deployed out of agriculture, thus increasing labour productivity in agriculture, thus increasing labour productivity in agriculture and providing labour for industry; and there is no doubt that a large percentage of the population in Italy did remain in the agricultural sector for far longer than in Britain, thus reducing overall productivity. Nor is it possible to contest the fact that, well after the process of industrialization was under way a surprisingly large proportion of national income was still produced in agriculture.

What is less clear is the precise significance of these features. The British paradigm would, of course, indicate that the persistence of the large agricultural sector would function as a brake on industrial progress. Yet, as we have seen, Heywood has argued against any automatic assumption in this respect where France is concerned. Unfortunately, in the Italian case, it has been all too easy to point to latifundia, absent landlords, small peasant farming and subsistence agriculture to support the orthodox view.

Recent studies suggest that this almost totally negative position requires, perhaps, some qualification. Purely at the level of agricultural efficiency, lengthy and exhaustive research has shown that only about 20 per cent of the total national agricultural product was consumed in subsistence agriculture.[20] The rest went to the market in one way or another. Similarly, a survey of sharecropping in Tuscany has demonstrated that the rationale behind the system was not purely that of social and political stability but also economic progress, and that this progress was consistently achieved.[21] And even the southern latifundia now have their defenders. Marta Petrusewicz's study of a large southern estate argues that the system of latifondo was socially and economically rational, at least up to the point of the agrarian crisis, and that the appearances of backwardness should not be confounded with a genuine degree of modernity. In response to generally held opinions, she writes that 'The picture which emerges [from the study] is so different from that painted by Salvemini that we can go back to almost all the old commonplaces of the "war on the latifondia" – rigidity, uniformity, monoculture and so on – and show that these are prejudices'.[22] At the very least, this might indicate that the usual division drawn between traditional and modern sectors is less than accurate for a good deal of Italian agriculture.[23] Areas which appeared backward by the logic of the large and efficient capitalist estates of the Po Valley were possibly less inefficient than assumed.

This may well be so but it is hardly the main point. The reassessment of the role of agriculture cannot seriously be posited on the idea that peasants were highly efficient producers or that the latifundia of the south were, within their own terms, rational organizations; whatever the necessary qualifications in this respect, to argue for an extremely efficient peasant agricultural sector would mean stretching the bounds of revisionist credibility just a bit too far. And it would obviously invite the question: if agriculture was doing so well, why did Italy have problems of accumulation at all? Any reassessment depends rather on recognizing that, in many areas, traditional agriculture played a very different role in the process of industrialization from that (passive, negative) of the productivity of labour. Nor can it be adequately summed up in the conventional distinction between backward, or traditional, and modern.

In fact, the nub of the question lies here. The dualism implicit in the traditional/modern divide is misleading if it is taken to mean that the traditional must necessarily give way to the modern for industrialization to move forward. This would, in effect, be simply a repetition of one aspect of the conventional paradigm of industrialization. It would mean that we are back to the search for large factories and monstrous machines to verify the very existence of the process.[24] The Italian case suggests instead that there is – for a very long period – an effective symbiosis between aspects of traditional and modern, particularly in rural areas, and that the specificity of the Italian example is constituted principally by this. Relations between families and their work – whether that work was agricultural or industrial – were frequently much more complex than is implied by recourse to any orthodox model of industrialization. The conventional passages from agriculture to industry, from peasant to proletarian worker, from country to town, were far less clear cut than in Britain – when they occurred at all, and often they did not. Indeed, as we shall see, the characteristic of large sections of the rural population is precisely the extent to which – despite increasing involvement with manufacturing – they avoided making those passages. The search for definition along the 'classic' lines of rural/urban and agricultural/industrial produces unsatisfactory results, therefore. Remarkably, lack of definition, rather than increased definition, seems for long to have been the characteristic of Italian transformation.

Peasant Families and Pluriactivity

Protagonist of this specifically Italian road to industrialization was the peasant-worker – or, more exactly, the peasant-worker family; for it is the development of a particular kind of family organization that determines the rather unusual aspects of Italian transformation. The best, but by no means the only, example of this comes from the silk industry, an industry that thrived in the area above Milan and which, by the middle of the nineteenth century, was producing raw silk for all the principal weavers of Europe.

The silk worms were reared by peasants – small, desperately poor, sharecroppers – who used the cocoons to pay off their debts, or a part of their debts, at the end of each year. The families, often multiple or extended, hated cocoon production because of the (relatively brief) campaign that required the voracious worms to be fed with mulberry leaves at all hours of day and night – to the extent that the worms themselves took over the house while the peasants made do in the outhouses. 'The worms kick men out of the best rooms' was a common observation of the time. Despite hardships, however, silk was the key to survival for the peasant families. This was for two reasons. First, it permitted peasants to keep some kind of hold on their land, through the payment of debts. More importantly, as silk-reeling mills moved out into the countryside to take advantage of wood, water, and, above all, cheap labour, peasant families found that, by sending the women and children to the mill, they could survive, and even possibly improve their situation.

In this way, during the second half of the nineteenth century, peasant families slowly adjusted to a situation of economic necessity in a way that saw the men involved in agriculture, and the women and children in manufacturing. This was the beginning of the 'pluriactive' family, in which different members of the family were occupied in either primary or secondary sectors. The families 'straddled' sectors, as it were, although – while the women would undoubtedly continue to work in the fields after the end of their factory work – the men remained firmly committed to agriculture. Pay from the mills was obviously very low. Employers imagined that peasant families drew most of their income from the land, and therefore gave what was only an 'integrative' wage to the women who worked for them.

Conversely, landowners increased their contractual pressure on the men, convinced that peasant families were making money in the mills.

In some ways, it might seem that we are back to protoindustrialization, with families involved in rural manufactures, dividing their time between the land and the mill. Alain Dewerpe, in his excellent studies on this subject, inclines towards a protoindustrial interpretation.[25] But, apart from the fact that diffused rural industry – which was the case of silk – is very different from domestic industry, the protoindustrial model hardly accounts for the behaviour of pluriactive peasant families *after* the first phase of contact with manufacturing. Protoindustrial theory would expect rural families eventually to become urban, proletarian paupers. What is remarkable in Italy is that this is exactly what does not happen. Resistance to these conditions remains the key feature of the behaviour of peasant-worker families.

The reasons for this behaviour were cultural and economic. Peasant families were intensely patriarchal in organization. They remained attached to the land because almost all decisions were taken by the men of the household, and these continued to express the values associated with the land. Here the hierarchy of gender operated in a conservative direction.[26] The men's objective – the pipedream of all sharecroppers – was to own the land they worked. Women's work in the mill did not disrupt this way of thinking. In fact, rather ironically, male attitudes were reinforced by the fact that women's wages permitted survival on the land. Thus, the sexual division of labour within the family did nothing to break down established hierarchies, quite the contrary.[27]

In economic terms, it is clear that the land did represent the possibility of a cheap house, of vegetables, chickens, rabbits, firewood, and so on. Many peasants also practised spare-time artisanal activities, and the land gave them space to do so. But the value of this resource can be appreciated only if it is put in relation to the alternatives. Urbanization was fraught with many unknowns for the rural population, but certain facts would have been common knowledge. At any time in the years between 1880 and the outbreak of World War I, immigrants to northern towns, such as Milan, were likely to find themselves in housing which was overcrowded, unhealthy and extremely expensive. Many of the enquiries carried out by the *Societa' umanitaria* in Milan in the first ten years of this century speak precisely of the very bad living conditions of urban workers, of the chronic shortage of housing, and of the high prices.[28] Rents for farm property were notoriously low when compared with town prices. This was, by itself, a very great incentive to remain in the farm house; rural hardship was still preferable to urban poverty.

The high cost of urban housing was compounded by the fact that immigrants from rural areas were likely to be among the worst-paid workers. Women could expect to find employment in textiles, garment making or personal domestic service – the three lowest-paid occupations. Men might fare better because building, the main employer of people without saleable skills, was better paid but it was seasonal and extremely insecure. High levels of unemployment were common among all workers, with building workers usually heading the list.[29]

Town life can have had few attractions for rural workers, therefore. Conditions were harsh for all, but recent immigrants could expect to be those who suffered the most in almost all respects. For women, certain of the disadvantages of employment in rural manufacturing were reproduced without compensating advantages; they were still paid very little because it was assumed that their income was a supplement to a male wage, and, because they tended to crowd into the same occupations, they were often in competition with one another. The areas in which they worked were also characterized by instability and rapid fluctuations in employment. Often, of course, the male wage was not forthcoming. And, even if workers could earn more in urban industries, quite clearly they also spent more. Not surprisingly, enquiries into the alimentation of workers' families found that they were often considerably below any reasonable level of sustenance.

But urban life imposed other problems as well. Small appartments compelled families to break up into nuclear units. This created difficulties of family organization when survival depended on all adults trying to work as much as possible. The larger families of the rural areas could always meet this difficulty; young children could generally be entrusted to grandparents who remained at home. In the town this was less likely to be the case. Although most children would be pushed into some form of employment at a very early age, there was inevitably a point in which the potential extra worker was a weight on the nuclear family.

Put in this way, the advantages of the peasant-worker household seem evident. The benefits of the house and the land, combined with the wage pooling of the multiple family, were likely to permit a much better standard of life than could be expected – at least at the outset – in the towns. It is not surprising, therefore, that permanent emigration to the town was seen as a loss of status for peasant families. Just as the initial entry of women into the factory had been considered demeaning to the family, so to leave the land on a permanent basis was seen to be a sign of defeat, of inability to make ends meet. On the other hand, commuting to work in the town, or temporary migration, was far more acceptable. Here it was possible to enjoy the benefits of the higher wages of the urban factories without having to pay the price of permanent residence or renouncing the advantages, of the rurally based family. Again, collective solidarity within the family seemed a better option than loss of contact with the land; and it must be emphasized that it was the strategy of pluriactivity that permitted such a choice.

Over the years, family organization around pluriactivity, originally imposed by necessity, became something of a habit, a mode of behaviour which there were few reasons to alter. Changes did take place but they did not alter the underlining family strategy. With the gradual decline of the silk industry and the growth of other opportunities for non-agricultural employment in the first decades of this century, men from peasant families – usually the sons – began to work full time in manufacturing; yet they continued to live in the farm house and to pool the wages with the others. In the immediate post-war period, savings were often used to buy small plots of land; sharecroppers finally became proprietors. But the fact that these plots were usually too small to support the family indicates fairly clearly that, even at the moment of purchase, the family was anticipating other sources of income from extra-agricultural labour. Increasing contact with industry inevitably put large families under certain strains. Habits changed and tastes altered. Marzio Barbagli describes families – now clearly more worker-peasant than peasant-worker – where the older men still ate black bread while the younger preferred white. The important point, though, is that they still sat at the same table to eat their different preferences.[30]

Inevitably, as the equilibrium between agricultural and industrial work shifted within the family, the old dominant values – those of the male peasant patriarch – began to lose their hold. The large, cheap house continued to be of great importance in the family economy, but the land was now clearly a convenient supplement to other, mainly non-agricultural, incomes. After World War I, a growing number of men worked in industry; male values changed accordingly and were reflected in new family aims. The old attachment to exclusively agricultural objectives began to disappear, to be replaced by increasing interest in manufacturing. But important elements of the old mentality remained. Members of worker-peasant families tended still to aim to achieve independence. A feature of the late 1920s and the 1930s is that of dependent workers employing the skills they had learned to set up on their own in some form of minuscule manufacturing activity. The sudden mushrooming of very small 'back-yard' businesses in many areas above Milan was the beginning of what, after World War II, would be a major industrial development. Such activities did not immediately preclude the continuation of pluriactivity, however. It is not difficult to imagine a multiple family in which there were introits from the land, from dependent labour in industry, and from independent manufacturing. Indeed, the greater the number of introits, the lower the risk attached to the initial leap into small-scale manufacturing.

The work-force which emerges (often hidden from the censuses) from this gradual evolution has many specific characteristics, therefore, and they are not those of the 'classic' paradigm of industrialization. It is a non-urban, non-proletarianized and not predominantly male work-force; it is extremely flexible, extremely attentive to economic opportunity, given to high levels of self-exploitation, not without some experience in manufacturing, and prepared to use initiative and take risks to achieve economic independence. It corresponds to an evolution typified more by adjustment to change than by radical breaches with traditional family organization, more by continuity than by cataclysm. The basic family strategy appears to have been the disposition to adopt almost any fresh expedient to avoid the break-up of the family itself, to embrace change to conserve fundamentals. As will be evident, the families which were moulded by what was, in the end, an enormously diversified and variegated experience, had very different attributes from the passive and dispirited labourers who left the land in England to become the 'low wage factory fodder' of the cities.[31]

Spontaneous and Molecular Development: Regions and 'Waves'

Those familiar with the recent literature on the Third Italy, flexible specialization, and industrial districts may find many familiar features in the above description. Dynamic small businesses, often family based, often growing out of the gradual dissolution of the sharecropping system, constitute the core of the most recent phenomenon.[32] Yet a careful examination of many rural areas of northern Italy suggests that the kind of development witnessed in the Third Italy in recent decades is not without historical precedents, at least in certain respects. The novelties of the present small-business boom certainly exist, determined in part by international trading conditions, but they should not obscure the fact that many of the same fundamental features of this development had appeared in other areas of northern and central Italy in earlier years. The similarities should not, perhaps, be pushed too far but the overriding impression is that of a series of 'waves' of often gradual, irregular and uncertain industrial development, involving much of the sub-Alpine belt, some areas of Emilia-Romagna and Tuscany, and, more recently, the Marches.

This was an industrialization that was 'grafted on' to slowly disintegrating agricultural structures, but which coexisted with those structures for a remarkably long period, served for some time even to maintain them, and which derived considerable benefits from them – not the least being low-cost labour. In these areas, the separation between agricultural and industrial employment was, for many decades, anything but total. It is interesting to note, for example, that, in the 1930s, full-time factory workers at the chemical complex at Porto Marghera would desert the factories at harvest time to help members of their families who still worked the land.[33] This was typical of the pluriactive family seeking, certainly, to exploit all economic opportunities, but attempting above all to avoid a situation in which agricultural and industrial employment became incompatible to the extent of compelling the break-up of the family.

The persistence of this intertwining of agricultural and industrial employment within the same family prompts several considerations. The first is that the traditional perspective on Italian industrialization is possibly somewhat distorted. The picture of a 'lazy' capitalism, very dependent on the State for protection and for orders, is undoubtedly correct for certain sectors of heavy industry. Yet the difficulties experienced by these sectors, and their slow growth, should not be considered the whole story. There now appear to be grounds for asserting that there was a different process in course – one which had begun earlier, but which continued in parallel with, and distinct from, the State-induced efforts. This was a 'spontaneous' and 'molecular' process, which was the result of a lengthy and very gradual social and economic transformation of rural areas, and on which State-promoted initiatives were eventually often superimposed.[34] It produced small-scale manufactures as the first visible signs of industrialization – some of which failed while some succeeded – but, above all, it produced a social humus of peasant-worker and peasant-artisan families particularly oriented towards enterprise and economic independence. This humus was a factor that found expression only very slowly in the unfavourable national and international trading conditions which Italy faced almost constantly before World War II, but which showed its value in the changed conditions of expansion of the second post-war period. In the sense that it was, in part, the result of a long period of incubation, the 'economic miracle' begins perhaps to look a little less miraculous.

Secondly, the identification of 'waves' of development tends to confirm the view of industrialization as a very long drawn-out phenomenon – as evolution rather than revolution – and as a highly regional process. The family strategies examined here had their origins more than 150 years ago. It is clear that much of northern Lombardy, as a part of the Habsburg Empire, was undergoing gradual transformation well before the unification of Italy, and that this was in large part a consequence of close contact with the main European markets. It is highly probable that Unification for a time interrupted, rather than furthered, this transformation. Indeed, the Lombard case rather points up the dangers of using aggregate national statistics to determine progress after 1860; disaggregated, the figures would undoubtedly reveal a region which had little to fear from a comparison with many other areas of industrializing Europe, and which certainly would hardly merit the generic term of 'late-comer'.

As outlined above, however, it was a regional industrialization with specific characteristics. The particular form of rural transformation, with a work-force that retained its contact with the land not despite, but often because of, involvement with manufactures, and which developed family strategies to accommodate the choices necessary to stay on the land, is most evident in northern Lombardy, even if more-or-less similar forms can be found at different periods in much of northern and central Italy. The diffusion of small and medium light industry in many formerly rural areas – a feature of contemporary Italy – would often appear to be linked to this social form of the pluriactive family. Where these family characteristics were not present, of course, was in the south, where forms of landholding and social habits rarely produced multiple and extended, still less pluriactive, families.[35] As a result, families organized on different lines; to begin with, the economic functions of women within the family were very different, precluding that kind of interchanging of roles which, in the end, made the northern pluriactive family so effective as an economic unit.

To argue for an element of 'spontaneity' in Italian industrialization is not so much to seek to overturn the existing picture of the 'late-comer', however, as to try to adjust that image. It is obviously important not to claim too much for the model of 'family' or 'spontaneous', as opposed to State-induced, industrialization; yet the identification of a more gradual process of industrial development around peasant-worker families and rural artisans would seem to suggest the need for a new emphasis: away from railways, banking and heavy industry and towards those rural areas which provided the context for industrialization. It is obviously not a question of substituting one set of subjects for another, but rather of attempting to find a new equilibrium between them. Gerschenkron's arguments may have occupied centre stage for too long; perhaps what was going on off-stage was just as, if not more, important.

In the same way, it has to be recognized, with Sabel and Zeitlin, that Manchester and Birmingham are bad masters if they continue to impose a view of industrialization based exclusively on factory and urban proletariat.[36] The social and behavioural consequences on sharecropping families of diffused rural industry, based on the exploitation of rural labour, seem to have been very different from those assumed by the 'classic' model of British industrialization, in which the maintenance of a high percentage of the population in agriculture would automatically be taken to be a cause of slow growth. Indeed, the study of the transformation of the peasant family in Italy suggests that, unless a broader approach to industrialization is employed – one which takes into account, not only factories and machinery, but also the less immediately obvious social and cultural changes among sectors of the population – important changes will remain unnoticed. Italian families developed attitudes and domestic strategies which were, in the long run, conducive to enterprise, given the right external conditions. The growth of a humus favourable to very high levels of economic activity and self-exploitation, usually linked to the desire for family autonomy and independence, was a contribution to industrialization which British agriculture was unable to provide. It represented the formation of a powerful and genuine 'enterprise culture' so often invoked in countries of now-declining industrialization.

Should we still consider agriculture as 'the ball of lead tied to the foot of Italian capitalism'? Possibly not. A longer view on industrialization prompts the suspicion that collective Italian expressions of guilt at the poor performance of the 'late-comer' may be misplaced. Rather, the Italian experience would seem to reinforce the view that peasants are not as antithetical to economic growth as once thought. Perhaps, therefore, in certain contexts, the persistence of a large agricultural population is not a 'deviation' from the true path of industrialization after all. Perhaps we need 'to recognize . . . that the economic, social and cultural foundations of an industrial capitalist order rest on much more than conventional measures of industrial or economic performance',[37] and to repeat once again that 'there was not one model for industrialisation in the nineteenth century – the British – but several'.[38] In the Italian case the slow evolution of the peasant family in certain regions seems to represent a distinctive and specific feature of industrialization that was not obviously abberrant or genetically flawed from the outset. This suggests, in turn, that the historical origins of Italian industrialization – like those of many other countries – may be much broader than the British paradigm has often led us to assume.

Notes

1 A. Gerschenkron, *Economic Backwardness in Historical Perspective* (Cambridge Mass., 1962).
2 E. Sereni, *Il capitalismo nelle campagne* (Torino, 1968) p. 146.
3 P. Grifone, *Il capitalismo finanziario in Italia* (Turin, 1970) *Introduzione, passim.*
4 R. Aldrich, 'Late-Comer or Early-Starter? New Views on French Economic History', *Journal of European Economic History*, 16, (1987) p. 89.
5 R. Roehl, 'French industrialisation: a reconsideration', *Explorations in Economic History*, 13, (1976) pp. 233–4.
6 A. Cobban, *A History of Modern France* (Harmondsworth, 1959) vol. 1, p. 1.
7 T. Kemp, *The Industrialisation of Nineteenth Century Europe* (Oxford, 1969) Chapters 2 and 3.
8 The debate developed around the work of R. Romeo, *Risorgimento e capitalismo* (Bari, 1959); a useful summary is to be found in the essay of D. Tosi in A. Caracciolo (ed.), *La formazione dell'Italia industriale* (Rome-Bari, 1969).

9 For a review of the revisionist positions *see* J. A. Davis, 'Industrialisation in Britain before 1850: New Prospectives and Old Problems', in P. Mathias and J. A. Davis, (eds), *The First Industrial Revolutions* (Oxford, 1989) pp. 44–68.
10 R. Cameron, 'A New View of European Industrialisation', *Economic History Review*, 34, (1985), p. 9.
11 N. F. R. Crafts, 'The New View of British Growth and Gerschenkron's Hypothesis', *Rivista di storia economica*, n.s., 6, (1989) p. 41.
12 Aldrich, 'Late-Comer or Early-Starter?', p. 89.
13 P. K. O'Brien, 'Do we have a typology for the study of European industrialisation in the XIXth century?', *Journal of European Economic History*, 15, (1986) p. 291.
14 Cameron, 'A New View', p. 23.
15 P. K. O'Brien, and C. Keyder, *Economic Growth in Britain and France 1780–1914* (London, 1978) p. 18.
16 *Ibid.*, p. 196.
17 C. Heywood, 'The Role of the Peasantry in French Industrialisation 1850–80', *English Historical Review*, 34, (1981) p. 376. *see also* Chapter 5 in this volume.
18 *See* in particular the works of Roehl, Cameron and Aldrich already cited.
19 R. Tilly, 'German Industrialisation and Gerschenkronian Backwardness', *Rivista di storia economica*, n.s., 6, (1989) p. 159.
20 G. Federico, '*Contadini e mercato: tattiche di sopravvivenza*' in *Società e storia*, (1987) 38, pp. 883–4.
21 *See*, for example, F. Galassi, '*Stasi e sviluppo nell'agricoltura toscana 1870–1914: primi risultati di uno studio aziendale*' in *Rivista di storia economica*, 3, (1986).
22 M. Petrusewicz, *Latifondo* (Venice, 1989) p. XIII.
23 This general conclusion is supported by many of the articles in the recent massive overall survey of Italian agriculture, P. Bevilacqua (ed.), *Storia dell'agricoltura italiana in eta' contemporanea*, 3 vols, (Venice, 1989–91).
24 M. Berg, 'Revisions and Revolutions: Technology and Productivity Change in Manufacture in Eighteenth Century England', in P. Mathias and J. A. Davis (eds), *Innovation and Technology in Europe* (Oxford, 1991) pp. 53–6.
25 A. Dewerpe, *L'industrie aux champs. Essai sur la protoindustrialisation en Italie du Nord (1800–1880)* (Rome, 1985).
26 Cf. the similar conclusions of G. Gullickson, *Spinners and Weavers of Auffay. Rural industry and the sexual division of labor in a French village, 1750–1850* (New York, 1986) and id., 'Love and Power in the Proto-industrial Family', in M. Berg (ed.), *Markets and Manufacture in Early Industrial Europe* (London and New York, 1985).

27 On the general question of the sexual division of labour in Italy, *see* F. Bettio, *The Sexual Division of Labour: The Italian Case* (Oxford, 1989) in particular Chapter 1.

28 R. Romani, *Un secolo di vita agricola in Lombardia (1861–1961)* (Milano, 1963) p. 159. *See also* G. Consonni and G. Tonon, '*Casa e lavoro nell'area milanese: dalla fine dell'ottocento all'avvento del fascismo*' in *Classe 14*, (1977).

29 For information specifically on Milan *see* L. Tilly, *Politics and Class in Milan 1881–1901* (New York, 1992) Chaps 3 and 4 in particular.

30 M. Barbagli, *Sotto lo stesso tetto. Mutamenti della famiglia in Italia dal XV al XX secolo* (Bologna, 1984).

31 The phrase is from R. Floud, 'Slow to grow', review of N. F. R. Crafts, *Times Literary Supplement* (19 July 1985), quoted in Berg, 'Revisions and Revolutions', p. 47.

32 There is a vast literature on this subject, but *see* in particular Bagnasco, A., *Tre Italie. La problematica territoriale dello sviluppo italiano* (Bologna, 1977), and id., *La costruzione sociale del mercato* (Bologna, 1988); M. Paci, *La struttura sociale italiana. Costante storiche e trasformazioni recenti* (Bologna, 1982), and S. Brusco, 'Productive Decentralisation and Social Integration: The Emilian Model', *Cambridge Journal of Economics*, 6, (1982).

33 F. Piva and G. Tattara (eds), *I primi operai di Porto Marghera: mercato, reclutamento, organizzazione 1917–40*, (Padua, 1983).

34 The terminology – spontaneous and molecular – is taken from L. Cafagna, *Dualismo e sviluppo nella storia d'Italia*, (Venice, 1989), *Introduzione, passim.*

35 Barbagli, *Sotto lo stesso tetto*, p. 119–20.

36 C. Sabel and J. Zeitlin, 'Historical alternatives to mass production: politics, markets and technology in nineteenth century industrialisation', *Past and Present*, CVIII, (1985).

37 M. Berg and P. Hudson, 'Rehabilitating the Industrial Revolution', *Economic History Review*, XLV, 1 (1992), p. 44.

38 Cameron, 'A New View', p. 9.

7

Agriculture and Industrialization: The Japanese Experience[1]

Kaoru Sugihara

The contribution of agriculture to Japan's industrialization from the second half of the nineteenth century on has been traditionally interpreted through the application of the models developed on the basis of Western (and, in particular, British) historical experiences. It has been argued that Japanese agriculture failed to transform itself into large-scale Western-style agriculture, and the low productivity of small-scale traditional agriculture forced the peasant household to suffer from the low standard of living. Thus, the low level of their purchasing power set the ceiling to the expansion of the domestic market. In contrast, the introduction of Western technology and organizations in the forms of steam engines, machinery and railways and steam ships, and of the factory system where the clock-time discipline dominated the pattern of work, have been regarded as a critical turning point. The relationship between industry and agriculture has often been understood to be that between *modern* industry and *traditional* agriculture, and the proportionate increase of modern industrial production in the gross national product has been seen as a sign of progress. The migration from rural to urban areas, and the industrialization accompanied by the urbanization, have been thought to be inevitable. Indeed, any divergence from this pattern, or even the slow progress in this direction, has been interpreted as an indication of Japan's relative backwardness (or 'lopsidedness').

This traditional historiography has assumed the availability of cheap labour, drawn from the countryside, to be a crucial factor for Japan's industrialization. After the Meiji Restoration of 1868 a strong central state was created and landlordism developed, and rural savings were transferred, through the land tax and the high rent, to industry for investment. In this way, agriculture provided industry with labour and with capital. The inter-war agricultural depression and the growth of rural-urban disparities led to the stagnation of the domestic market, and revealed the limits of this State-led and cheap, labour-based industrialization strategy. The export drive and expansionism in the 1930s were the direct result of this 'impasse' of Japanese capitalism.[2]

In the last 20 years or so, economic historians have challenged this traditional view, revising our understanding of some aspects of living standards upwards, describing the process of innovative adaptations of Western technology and organizations in detail, and making the general picture of the economy a much more dynamic one.[3] A most fundamental revision has been a new emphasis on the positive contributions of indigenous elements (though a lot of them came originally from China) to agricultural development and industrialization. The discovery of 'high' initial conditions before the Western impact led to the more serious examination of the economic rationale behind indigenous social institutions, small-scale agriculture and protoindustrialization. The effects of the opening of Japanese ports to foreign trade in 1859 and the Meiji Restoration remain central themes, but they are no longer the obvious starting points of economic modernization. The focus of analysis on the second half of the nineteenth century has shifted from the wholesale and direct introduction of Western technology and organizations to the more diverse and indirect, but no less significant, processes of absorption of foreign technology and organizations into the domestic economic and social contexts. Given the nature of domestic and Asian demand, a sensible strategy for increasing output and exports was to improve the production methods of traditional commodities, such as rice and kimono clothes, rather than to compete with Western countries for products requiring top technology. The main policy goal was to activate and modernize the existing rural economic structure by providing the essential support, such as education, market information, modern transport and energy.

To some extent, these revisions corresponded to the revised understanding of British and European experiences.[4] But they also questioned the conventional methods of comparing Japan with other countries. In some respects, the Japanese experience was vastly different from Western European norms. Farm size, for example, was much smaller; About 70 per cent of Japanese farms had 0.5 hectare of land or less, and nearly 90 per cent 1 hectare or less at the time of industrialization. Per capita income was also significantly lower although other aspects of living standards, such as education and hygiene were comparatively high. More relevant comparisons might be made among the Asian experiences. Japan shared so many common features with other

Asian countries, and the historical circumstances under which Japan responded to the Western encroachment from the middle of the nineteenth century were also similar to those of neighbouring countries. The Japanese experience might be better assessed if we abandon the assumption of unilinear (and often Eurocentric) patterns of development. According to this line of thought, even the difficulties in inter-war years, which was the starting point of the traditional view, may not necessarily represent the limits of the Japanese pattern of development. Behind the politically and militarily disastrous decisions in the 1930s lay a fundamentally international question of how to secure Japan's economic interests during the world depression? The failure to deliver the answer to this question may suggest that Japan's industrialization strategy lacked a sound international, economic and political management programme. But it does not prove that the more wholesale urbanization and westernization would have provided an answer.

This chapter offers a summary interpretation of current scholarship on the relationship between agriculture and industrialization, broadly from a perspective sympathetic to this latter view. Attempts are made to reinterpret some of the key observations of the traditional view, however, particularly on labour and the role of indigenous social institutions, and incorporate them into the revised picture. The primary aim is to highlight the rural nature of Japan's industrialization in the late nineteenth and the early twentieth centuries, and locate it in a long-term historical context.

The Tokugawa Heritage

During the Tokugawa period (1603–1868), the Tokugawa house ruled Japan from Edo (present-day Tokyo), controlling about a quarter of land directly and assigning the rest to those domainal lords who swore allegiance. The ruling class (*samurai*), merchants and artisans were separated from agricultural production and lived in major cities, castle towns and local towns, while the countryside was divided into small villages where farmers, an overwhelming majority of the population, resided. The village was responsible for the collection and payment of tax, normally paid in rice, in exchange for which it enjoyed a considerable degree of political autonomy. The typical daily life of the village in this period was a relatively peaceful one, disturbed only by poor harvest and occasional famines. Farmers were, however, tied to the land, and the production of rice, not of a cash crop of their choice, was expected. Marriage with non-farmers was prohibited in principle. Except towards the end of the period, village boundaries were fairly strictly observed, so that rich farmers could accumulate land in their village only.

This severely limited the development of land market and capital accumulation, hence the emergence of rural entrepreneurs.

In terms of crops, Japanese agriculture increasingly came to resemble Chinese agriculture during the seventeenth century. Local grain and hemp production declined, and rice, along with cotton and silk, began to dominate the Japanese countryside. The population grew very fast (the annual rates of growth are estimated to have been 0.72 per cent in 1600–50 and 0.96 per cent in 1650–1700 respectively), which was accompanied by the growth of arable land and the increase of rice output. Active international exchange in the second half of the sixteenth and seventeenth centuries was responsible for this transformation. For example, the introduction of comparatively drought- and flood-resistant red Indica rice varieties made it easier to open up the fan-shaped delta near the mouth of the main river to turn it into arable land.

By the early eighteenth century such possibilities were exhausted and foreign trade declined, and a strong demographic pressure on land occurred. The Japanese village responded to this by increasing land productivity and by controlling population. The central device here was the tightening of the village community and the household (*ie*) system. The household became a quasi-kinship organization (adoption was frequent), in which the maintenance of the family line was assumed to be of primary importance. The status of the household in the village was also vital in so far as its successful maintenance was measured mainly in this context. The village and the household thus became the two key social institutions under which economic activities were organized and general goals set out.[5] The economic development and demographic change under this institutional framework formed the basis of modern Japanese agriculture.

The most immediate response to this demographic pressure was to intensify land use by absorbing more family labour. The development, commonly associated with labour-intensive technology, included the use of different seed varieties (early- and late-ripening rice varieties, for example), double cropping, the use of manure (dried fish, oil cakes and night soil), and the development of agricultural tools and small irrigation works. Agricultural manuals in the late seventeenth century onwards concentrated on the promotion of commercial crops, better rice seed varieties and suitable agricultural tools. They were based largely on the Chinese agricultural manuals, but also contained detailed local knowledge. These developments were interrelated. The provision of a good drainage system made it possible to change wet rice land into dry land for winter crops in autumn, and back to wet land for rice cultivation in spring. In advanced regions, this allowed the diffusion of dry-field horse-ploughing which ensured good soil conditions. Double cropping involved the absorption of off-peak family labour as

Table 7.1 Estimates of pre-industrial rice yields in Japan and other Asian countries

	Paddy rice yield (tonne/ ha)
Japan 1878–82	2.53
other Asian countries (FAO figures 1953–62)	
India	1.36
Thailand	1.38
Indonesia	1.74
Malaya	2.24
Korea	2.75
Taiwan	2.93

Source: Yūjirō Hayami and Saburō Yamada, 'Agricultural Productivity at the Beginning of Industrialization', in K. Ohkawa, B. Johnston and H. Kaneda (eds), *Agriculture and Economic Growth: Japan's Experience*, (Tokyo, 1969) p. 108.
Note: The above Japanese figure is the revised official estimate. Other estimates range from 2.36 to 3.22 tonnes/ha.

well as the development of labour-sharing practices during peak seasons. The use of female and child labour was made possible partly by the development of a variety of simple iron tools. Technological advance of this type was based on the more intensive use of the existing factors of production, land and labour, with a relatively small input of capital and a relatively limited degree of market involvement. Around 1730, per capita agricultural output stopped declining and began to rise continuously thereafter, although labour productivity may have stagnated or even declined if greater labour inputs per capita were taken into account. But land productivity rose steadily. The annual rate of increase for 1730–1800 is estimated to have been 0.19 per cent and 0.09 per cent for 1800–50. The importance of this development can be seen when Japan's land productivity (rice output per area) at the beginning of the Meiji period is compared with that of other Asian rice economies at a similar stage of economic development (*see* Table 7.1).[6]

The village and the household also functioned as the management unit of population under the pressure on land. In the 1734 official survey, for example, the typical household illustrated in the 'model village' is assumed to have cultivated only about a hectare of land. There was now very little chance for the subdivision of land. It became increasingly difficult to get a new household 'approved' in the village, and, even if it was approved, its status was likely to be inferior to that of the existing household. The status mattered, not just in village politics and ritual rights, but in the allocation of water and the sharing of labour. Thus, there were good reasons for 'family planning' through infanticide and abortion. The former implied sex selection (in favour of males) as well as the control of the number of children. Some economic historians suggested that this was the result of farmers' conscious attempts to raise their standard of living. But infanticide and abortion alone are unlikely to explain the low 'birth rate'. In some cases, marital fertility itself was lower than the natural level, despite the fact that the average calorific intake was probably adequate. It is possible that the development of labour-intensive technology meant that in the eighteenth century, women worked harder during their pregnancy, contributing to lower fertility. If that is the case, the picture may have been a harsher one. The poor peasant household may have produced fewer children in any case.[7] Whichever the case may have been, Tokugawa demographic history lacked drastic Malthusian checks on a nationwide scale. Although there were some famines, catastrophies, such as epidemics and warfare, played little part in the secular trend, and mortality remained relatively low. Yet the Japanese population largely stagnated between 1721 and 1846 at a little over 30 million. This stability was crucial for establishing 'high initial conditions' at the time of Western impact; greater labour absorption without more mouths to feed was the key to the Tokugawa development.

After the second half of the eighteenth century, major urban centres and castle towns declined, while rural industries began to grow. Rural merchants engaged in intraregional commerce, while feudal domains actively pursued policies to promote agriculture, commerce and industry to earn 'foreign' exchange. Both of these activities gave farmers a chance to exploit non-agricultural, as well as agricultural, economic opportunities. The rural household mobilized cheap labour, responding to the fragmental rise in rural purchasing power. By the end of the eighteenth century, the daughter of a rich farmer is likely to have included a silk kimono in her dowry, but this did not have to be produced in the city of Kyoto where the most elaborate kimonos were made. Interregional merchants could also cut the margins enjoyed by the merchant guilds in Osaka and Edo.

From the point of view of the rural household, this was merely an extension of their labour-absorption strategy which had been developed in small-scale agriculture. For example, the rural merchant would bring a loom and yarn to the peasant household and collect the cloth a month later, thus providing a small amount of income for the housewife-cum-weaver. Some cottage industries put workers into one place to manufacture sake, using simple tools and water power. For the rural household, the 'main' agricultural work was rice cultivation. Non-rice cash-crop production as well as protoindustrial work of all sorts were called 'additional' work, be it by-employment at home or hired labour. It was important that rice production, the tax base of the Tokugawa regime, was recognized as a priority.

In the West, there was a tendency for protoindustrial regions to specialize geographically in industries and develop trade with regions with commercialized agriculture. In Japan, the division of labour between agriculture and industry occurred through the allocation of family labour, particularly in the form of sexual division of labour. The 'main' agricultural work was considered to be primarily a man's job, while women engaged in 'subsidiary' agricultural work as well as protoindustrial employment, particularly of silk reeling and cotton weaving.[8] There was relatively little need for urban growth and rural-urban migration. The absorption of rural female labour implied the relative decline of urban industry. The growth of demand for female employment in the rural silk and cotton industries also visibly corrected the biased sex ratio in protoindustrial regions. Population control was guided by the social norm of the village community and the will to maintain the family line, and these guidelines directed the household to respond sensitively to economic change at a micro-level. There was little room for the emergence of a class of agricultural labourers here.

For most of the period, the village community was able to prevent tax increases, which meant that the increase of agricultural and non-agricultural output resulted in the accumulation of surplus in the village and the decline of the effective tax burden. The latter probably declined from about 50 per cent in the seventeenth century to less than 30 per cent in the early 1870s. This was another reason for urban decline, and also an important background to the fall of the Tokugawa regime. Around 1820, a long-term trend of inflation was set in motion, due initially to the recoinage which was designed to counter the regime's fiscal crisis. As the increase of wage rates lagged behind inflation, rural entrepreneurs expanded their activities and accumulated their wealth, ignoring the rules of occupational division and village boundaries. Some of them were behind the Meiji Restoration of 1868, initiated by lower-or middle-class *samurai* in south-western Japan.

The Meiji Agriculture

The Meiji government removed many feudal restrictions, approving of freedom of occupation, migration, the sale of land and the free choice of crops for farmers. Building up Western-style military and industrial strength was vigorously attempted. The government initially financed the development of arsenals, shipyards and model factories of silk reeling and cotton spinning, employed foreign engineers, and sent Japanese students abroad to acquire Western technological and organizational skills. Seen as a learning process, these initiatives were extremely important. They provided entrepreneurs with valuable information and useful lessons. Some of the workers left government enterprises after a period, and started their own businesses, using the skills they acquired in their previous employment.

Initially, Japan was able to finance the import of manufactured goods from the West through the outflow of bullion, rather than through capital imports, to effect this modernization programme. During the deflation of the early 1880s, the balance-of-payments crisis triggered a serious review of the programme, and many government enterprises, including model factories and mines, were sold off. In 1884, *Proposals for Enterprise Promotion* (*Kōgyō Iken*), the first comprehensive economic policy document, was compiled at the Ministry of Agricultural and Commercial Affairs (MACA), and the focus was placed firmly on the promotion of agriculture and traditional industry rather than on the direct introduction of Western technology and organization.[9] Not only were raw silk, tea and rice the main earners of foreign exchange, but agriculture provided the State with its firm fiscal base. Between 1873 and 1876, the land-tax revision was carried out, and a centralized modern taxation system was created. The tax was now to be collected in cash rather than in kind, and, after the revision, about 70 per cent of the fiscal revenue came from the land tax. The revenue from those regions which benefited from exports of silk and tea was crucial at this stage.

The government's initial attempts to introduce Western-style agriculture largely failed, because imported machinery and foreign plants and animals were not easily compatible with small-scale agriculture and different climatic and soil conditions. Discontinuing most of the early experimental stations and facilities, MACA implemented measures to improve and diffuse indigenous agricultural technology. The Experimental Farm for Staple Cereals and Vegetables (established in 1885) tested seed varieties and practices, and many veteran farmers were employed as instructors, together with the graduates of agricultural colleges, to combine practical knowledge with modern science. The Farm developed into the National Agricultural Experimental Station in 1893. The Itinerant Instructor System was created in 1885 to diffuse better seed varieties and practices. After 1899 many local governments set up prefectural agricultural experimental stations with central government subsidies.[10]

In this process, rich farmers and cultivating landlords emerged as the main agents of innovation as well as the beneficiaries of State initiatives. With the development of the land market, the tenancy rate increased from some 30 per cent in 1872 to 40 per cent by 1887, and to 45 per cent by 1912. The rate of population growth was not particularly high, but, given the scarcity of land and the reluctance to emigrate, there was no prospect for the improvement of the land-labour ratio. Hence, the rent remained very high (about half of the total production) in the Meiji period. Meanwhile, the general trend of inflation and the government's inability to revise the official value of land upwards meant that the proportion of land tax to the value of agricultural production went down further to around 10 per cent by the late 1870s, and remained at that level until the early twentieth century. The result was that the owners of land, landlords and owner-cultivators alike, gained while an increased number of tenant farmers were left out from this change in income redistribution. Part of the surplus accumulated by landlords and owner-cultivators was channelled into industry. But they invested the bulk of it in agriculture, especially the improvements in land infrastructure such as irrigation and drainage.[11]

The key improvement centred around the seed-fertilizer technology. Selected rice varieties were recommended by veteran farmers and widely publicized by the government. The extensive use of commercial fertilizers was made possible by the introduction of cheap commercial fertilizers and the more fertilizer-responsive rice varieties. Improved transport and distribution networks helped, not only the development of the market for agricultural products, but the diffusion of commercial fertilizers. The development of seed-fertilizer technology was based on the land infrastructure developed in the Tokugawa period, but it also created a justification for its further improvement. The government took a series of measures to facilitate projects for land improvement, including the establishment of Japan Hypothec Bank in 1897 and regional agriculture and industry banks. No less important was the development of agricultural co-operatives which provided small farmers with short-term credit. They were organized on the basis of the village community, and were instrumental in rescuing run-down villages and resisting the transfer of land ownership to outsiders. The village also took on a new organizer role by forming marketing co-operatives. Although initially developed voluntarily, the government encouraged the establishment of these credit and marketing co-operatives, and was responsible for its rapid diffusion.[12]

In spite of the increased importance of investment, therefore, Meiji agriculture firmly followed the Tokugawa pattern of development which was biased towards labour-intensive technology. Multiple cropping and dry horse-ploughing diffused further to accommodate greater inputs of labour and fertilizer. For example, the development of summer–autumn rearing of cocoons enabled farmers to combine rice production with sericulture, because, unlike the spring–summer rearing, it avoided the peak season of farm labour overlapping with that of rice.[13] Development of this kind implied an increase of working hours per worker-year. Also, female labour became more heavily drawn into agriculture. Greater labour absorption probably accounted for the majority of the increase in agricultural output in the Meiji period. Compared to the previous generation, members of the typical late-Meiji household enjoyed a higher income, better education and managed more sophisticated production patterns. But they also worked longer hours.

Rural stratification between the owners of land and tenant farmers did occur, and it was the former category that took technological and managerial initiatives. But the consolidation of land ownership was not meant to disturb the role of small management units, particularly their capacity to absorb labour. The concentration of ownership had its own limits, and the tenancy rate stayed between 40 and 50 per cent for most of the pre-World War II period. Thus, small-scale agriculture and the village-based social and economic organizations remained the backbone of the Meiji economy. In 1881–85, 71 per cent of the population was engaged in agriculture, and the figure was still as high as 61 per cent in 1911–15, although the share of primary-sector production in Net Domestic Product (NDP) fell from 45 per cent in 1885 to 28 per cent in 1915 (*See also* Table 7.2).

Table 7.2 Agriculture in Japan's economic development, 1885–1980 (per cent unless otherwise stated)

	Real GDP per capita (in 1980 dollars)	*Share of agriculture*		*Agriculture industry labour productivity ratio (1)*	*Farm/ non-farm household income ratio (2)*	*Agri./ manufacturing terms of trade (1885 = 100)*
		Labour force	*GDP*			
1885	630	73	45	75	76	100
1890	710	71	48	67	87	115
1900	880	68	39	49	52	102
1910	1000	65	32	37	47	98
1920	1260	54	30	50	48	99
1930	1350	50	18	31	32	104
1935	1620	47	18	24	38	136
1955	1850	39	21	55	77	163
1960	2690	32	13	39	68	169
1970	6920	17	7	25	91	304
1980	9890	10	4	17	115	347

Source: Yūjirō Hayami, *Japanese Agriculture under Siege: The Political Economy of Agricultural Policies* (London, 1988), pp. 20–1.

Notes: 1 The ratio of real GDP per worker in agriculture, forestry and fishery to real GDP per worker in mining and manufacturing.

2 The ratio of average income per family member in farm household to that of urban-worker households.

Traditional Industry and Modern Industry

The share of the secondary (industrial) sector production in NDP rose from 15 per cent in 1885 to 32 per cent in 1915, while that of the tertiary (service) sector remained at 40 per cent in the same period. If we call the industry engaged in the production and distribution of goods and services which had existed in the Tokugawa period, 'traditional industry', and distinguish it from 'modern industry', an overwhelming proportion of the Meiji economy consisted of traditional industry. Not only did most of agriculture and services fall into this category, but the large traditional industrial sector continued to grow in size. The modern industrial sector grew more rapidly, but its absolute size in terms of output and employment was not yet dominant. The concurrent growth of agriculture, traditional industry (hereafter, refers to manufacturing only) and modern industry was thus a major characteristic of the Meiji industrialization.[14]

The typical traditional industry was located in rural areas, in the form of either family (or small) business or farm family by-employment organized by merchants. The 1909 Factory Survey reported that 72 per cent of 'factories' (defined as a concern employing five or more persons) had no power supply, and 5 per cent operated on the Japanese-style (small) water mill, while the factories equipped with steam engines, gas-or petrol-operated engines or motors, consisted of 23 per cent. Some 58 per cent of 'factory' workers were employed in 'small factories' (employing between five and 100 persons). In addition, production employing less than five persons, which was excluded in the Survey, is estimated to have amounted to 51 per cent of total industrial production.

Traditional industry was dominated by food processing (particularly brewing of soysauce, bean paste and sake) and textile-related sectors (particularly silk reeling, cotton weaving and indigo production). Unlike the former, the latter sector was severely affected by the opening of Japanese ports to foreign trade, and experienced drastic structural changes in production and demand. The two main branches, silk and cotton, played a central role in the concurrent growth of agriculture, traditional industry and modern industry. Sericulture and silk reeling grew hand in hand, while cotton developed the largest modern sector, along with the growth of hand-loom weaving, and helped the growth of the infant machinery sector. We will sketch below these two developments and their strong rural orientation.

The opening of ports gave sericulture and silk-reeling regions in central and eastern Japan great market opportunities, and the export of raw silk to the United States and Europe became the single largest foreign-exchange earner for most of the pre-World War II period. The Meiji State provided

Table 7.3 Raw silk production and exports, 1818–1900 (1000 tons)

Production		*Exports (less imports)*	*Domestic consumption*
1868–73	n.a.	0.60	n.a.
1874–75	2.31	0.66 (28)	1.65 (72)
1876–80	2.83	0.99 (35)	1.84 (65)
1881–85	3.52	1.50 (43)	2.02 (57)
1886–90	3.52	2.01 (57)	1.51 (43)
1891–95	5.09	3.03 (60)	2.01 (40)
1896–1900	6.47	3.11 (48)	3.35 (52)

Source: Takafusa Nakamura, *Meiji Taishōki no Keizai (The Economy of the Meiji and the Taishō Period)*, (Tokyo, 1985), p. 243.
Note: Production figures are Nakamura's estimates based on silk-worm production statistics, and are substantially higher than the contemporary raw silk production figures by MACA. The percentages are based on the original, rather than rounded, figures.

export merchants with critical credit facilities, while the treaty port system acted in their favour because it prevented foreign merchants from penetrating into the domestic distribution system. Japanese silk competed well with Chinese silk in the United States' market of low-quality silk from the late 1880s partly because, in the lower range of the market, the quality of Japanese raw silk was better controlled as a result of government intervention and guidance. The other less well-known factor was the potential competition between British cotton textiles and silk manufacture in the domestic market. Fine cotton cloth could be a substitute for low-quality silk cloth. But, while Nishijin, the traditional centre for silk manufacture in Kyoto, was severely hit by the high price of raw silk resulting from the export boom, the more ordinary silk kimonos began to be manufactured in much larger quantities by rural weaving centres. Combining the more efficient hand-loom with cheap labour, they captured the growing domestic market and prevented the influx of foreign textiles. As Table 7.3 shows, the increase in the domestic consumption of raw silk was as important as that of exports during the nineteenth century.

Central to this development of silk-reeling industry were the introduction of simplified spinning machines and the diffusion of improved sedentary reeling machines. The government built model factories in the early 1870s, with the support of foreign engineers and French and Italian technology, but the imported machines did not spread quickly. Instead, simplified spinning machines which were cheaper and better suited to the production of lower range silk, were devised in Japan, and they were extensively used for production for export. At the same time, the improved sedentary reeling machines diffused rapidly, replacing traditional hand-reeling machines. Domestic wholesalers and other merchants often loaned them or offered credit to small-scale producers. Though initially a vital export, too, rural weaving centres consumed most of the raw silk produced by this method as well as some from spinning machines. Sedentary reeling remained important until the changes in the quality of overseas demand and the introduction of power-looms in Japan began to favour spinning machines in the second half of the 1900s.[15]

Perhaps the most important result of the development of the silk industry was its effects on employment. In the early twentieth century, over 100,000 female workers were employed in about 3500 factories, while over 160,000 workers operated sedentary reeling machines for about 10,000 small businesses. Further, both of them consumed silkworms which were produced by about 880,000 farm households where about 1.51 million people were engaged in sericulture.[16] This last figure suggests the critical importance of the summer–autumn rearing technology mentioned above which made the combination of rice and silkworm production so much easier. Sericulture became the main, or primary, supplementary crop in central and eastern Japan. The technological development of small-scale agriculture and that of traditional industry reinforced each other in this way.

The effects of the opening of ports on the cotton industry were more complex, but its response was essentially similar to that of the silk industry. The influx of British cotton textiles led initially to the decline of some local weaving centres, but many other centres responded to the competition by switching its raw material from hand-spun yarn to imported British and Indian yarn, while some new centres emerged (*see* Table 7.4). Part of the reasons for this competitiveness lay in the difference in consumer taste. The Japanese were used to traditional thick cloth made of coarse yarn, which did not directly compete with finer British textiles. The narrow size of kimono cloth also favoured domestic producers. But it was also the result of the successful use of machine-spun yarn by traditional weavers. Local weaving centres introduced the more efficient hand-loom (*takabata*) which had been used only for silk manufacture in the Tokugawa period. Unlike hand-spun yarn, foreign (machine-spun) yarn was strong enough to stand the tension which was put on it in the *takabata*-type loom when it was used as warp. Thus, some weavers used foreign yarn as warp and hand-spun yarn as weft. The fact that all the traditional weaving centres, which had declined at this time, failed to introduce *takabata* and foreign yarn implies that the introduction of *takabata* to cotton weaving was crucial for this development. Wholesale merchants were often responsible for expanding local weaving centres by loaning the improved hand-looms to rural household weavers.[17]

Table 7.4 Cotton textiles imports and production, 1861–1900 (1000 kin = 60 tons)

	Imports	*Domestic production*			
		Total	*with imported yarn*	*with hand-spun yarn*	*with machine-spun yarn*
1861	31	281	3	278	
1867	78	167	23	142	2
1874	173	256	115	134	7
1880	182	597	315	271	11
1883	104	451	271	141	39
1888	157	899	522	271	106
1891	117	917	191	198	528
1897	209	1569	177	167	1225
1900	313	1538	100		1438

Source: Satoru Nakamura, *Meiji Ishin no Kiso* Kōzō (*Structural Foundations of the Meiji Restoration*) (Tokyo, 1968), p. 221.

Domestic raw-cotton production and hand-spinning declined almost entirely as a result, which affected the rural household in western Japan. On the other hand, this development gave a great opportunity for the emergence of the modern spinning industry. Early government mills were ill-conceived and financially unsuccessful but the success of the Osaka Spinning Company, which started production in 1883 with a mule of more than 10,000 spindles, demonstatrated the economic viability of modern factory operation. A few Japanese engineers were able to produce 15-to 20-count yarn without the presence of foreign engineers, which was the right (low) count for the domestic market. Following the success of this company, many mills were established in the late 1880s. The ring frame, which was new and suited for the production of low-count yarn, was imported through Mitsui Bussan general trading company from Platt Brothers, and rapidly diffused. Short-staple cotton, suited for the production of low-count yarn, was imported initially from China but, in the 1890s, direct links with Indian producers were established to secure stable cotton supply. The Cotton Spinners' Association was formed partly to press the government to lift the import duty on raw cotton and provide freight subsidies for the import of cotton from India. In the 1890s, Japanese mills enjoyed extremely favourable conditions for exports when Indian exports of cotton yarn to China disturbed by were the adoption of the gold standard in British India and the consequent rise of the value of the rupee against silver-linked tael and yen. In the 1900s, the ingenious technique of mixing short-staple Indian cotton with a small amount of long-staple American cotton was developed in Japan to save costs and also to shift gradually production towards the slightly higher-count yarn. Some mills began to set up their own weaving sector, while the demand for improved hand-looms (and eventually power-looms) provided the basis for the development of a machinary sector.[18]

An overriding concern in this process was to minimize the cost of capital, which was scarce. The introduction of foreign machinery was thus accompanied by a variety of capital-saving devices. Together with the spread of the ring frame, which was relatively simple to operate, young country girls of 15 to 20 years of age were recruited from poor peasant households of relatively distant places, and were put into dormitories during their stay (normally two to three years) as factory workers. The industry was able to save wage costs by selecting this section of the labour-force which was expected to play a peripheral role in the maintenance and reproduction of the rural household. This was an effective way of recruiting and managing labour, although long working hours, harsh working conditions and the prevalence of tuberculosis caused much concern. The dormitory also suited the night-shift system, which was another capital-saving device. To the extent that Japanese agriculture was labour-intensive, these girls were used to hard work and long working hours. This gave Japanese mills a distinct advantage over the competing Indian mills which suffered from lack of discipline among their workers. Japanese workers accepted rural social values of loyalty and devotion, and the knowledge that their performance in the factory would be reported to their parents and to the village community at large, not only prevented them from running away from the factory when working conditions were harsh, but motivated them to compete with fellow workers for the place of 'model worker'. Japanese mills took advantage of this strong rural societal base, and attempted to build on these traditional values to establish their authority.[19]

It has been suggested, in line with the Gerschenkronian model, that the main goal of the Meiji State was to narrow the large technological gap between Japan and the West. But the more relevant question was how Western technology and organization could be absorbed into a non-European economy with a very different commodity and technology mix and institutional framework.[20] Thus, the Meiji government was much more concerned with the exploitation of rural human resources and their technical and managerial knowledge. During the latter half of the Meiji period, a large amount of local and central government expenditure went into the establishment of commercial and technical schools, commercial museums and industrial experimental and testing stations. The Ministry, of Foreign Affairs and MACA were eager to provide small businesses with overseas commercial information, while MACA took measures to improve standardization techniques for mass production. The conscious exposure of relevant information, government-sponsored exhibitions and the public acknowledgement of successful inventions encouraged keen domestic competition, while the priority areas for development were identified, and rules for competition were set out by the government; certain malpractices were checked and controlled through the regulations on local trade associations. Thus, a nationwide system of regulated competition emerged in this period. It formed the basis of modern Japanese industrial policy.[21]

Inter-war Developments and Urbanization

In 1920, 22.8 million people, or 84 per cent of the total population employed, worked in the countryside, of which 14.9 million were in the primary sector (mostly agriculture), 3.6 million in the secondary sector (mostly manufacturing), and 4.3 million in the tertiary sector (mostly commerce and other services). By contrast, only 4.4 million, or 16 per cent, worked in cities with a population of more than 20,000, of which 1.7 million were in the secondary sector and 2.5 million in the tertiary sector. The countryside provided the main source of employment, not only for agriculture, but also for industry. It was also the main source of employment for the modern manufacturing sector. This sector employed 1.1 million in the countryside and 0.6 million in cities.[22]

In addition, the bulk of workers in the modern manufacturing sector consisted of temporary migrants, such as young factory girls. The rural household thus remained the only major societal base under which industrial labour was produced. It is therefore inappropriate to apply a one-way labour-migration model to the Japanese experience in the way that the 'dual economy' theory did.[23] The rural household was in a position to decide the allocation of labour between agriculture and industry, and gradually increased the latter, while retaining its social values and economic goals. As Table 7.2 shows, the ratio of agricultural labour productivity to industrial labour productivity moved roughly in accord with the ratio of farm household income to non-farm household income before 1930, reflecting the relatively constant domestic terms of trade between agriculture and industry.

During World War I, the industrial boom created great demand for labour and led to high wages. A small core of skilled workers in heavy industry began to form urban households with income sufficient to support their families. Thus, it has been argued that Japan reached Lewis's 'turning point' (from the labour-surplus to the labour-shortage economy). But the inter-war years also saw the shift towards agricultural protection, and the domestic terms of trade began to favour agriculture (*see* Table 7.2). In the 1930s, stagnant industrial wages were accompanied by a rise on agricultural prices. This growth of income of the rural household prevented rapid urbanization, and offered industry an expanding domestic market as well as a release of labour at competitive rates. With imports of food and raw materials from, and exports of industrial goods to, other Asian countries, Japan achieved the fastest industrial recovery from the world depression among the industrial nations.[24]

Thus, between the wars, the rural household economy was only slowly transformed into the urban household economy, and traditional small-scale production was only slowly transformed into modern large-scale production. After the disruption of World War II, the process of urbanization accelerated and the rise of big business became apparent in the 1950s and 1960s. The proportion of city dwellers in the total population rose from 38 per cent in 1950 to 76 per cent in 1975. The Japanese economy shifted its base from the rural household to the urban household at this point, and a persistent rise in wages resulted. To maintain the quality of labour with reasonable wage costs, it was necessary to form the stable urban household quickly and smoothly, and big-business management sought to respond to the needs of urban workers and their families. The spread of lifetime employment, seniority wages, welfare facilities, occupational pensions and 'companyism' as an ideology all helped to fill the gap created by the rapid disappearance of the rural household and the village community. It should be remembered that a large part of these ideas came directly from the experience of rurally based industrialization before World War II.

Notes

1 Penelope Francks and Osamu Saitō read the earlier versions of this paper and made valuable comments. The usual disclaimer applies. In this paper references are confined to English-language literature as far as possible. The descriptive evidence and data cited here without references can be found in either Kazushi Ohkawa, Miyohei Shinohara and Mataji Umemura (eds,) *Estimates of Longterm Economic Statistics of Japan since 1968*, 14 vols, (Tokyo, 1965–88) or Mataji Umemura *et al.* (eds,) *Nihon Keizaishi* (*The Economic History of Japan*), 8 vols, (Tokyo, 1988–90). Japanese names appear in western order, i.e. first name followed by family name.

2 For a summary of this traditional view, *see* Kaoru Sugihara, 'The Japanese Capitalism Debate, 1927–1937', in Peter Robb (ed.), *Agrarian Structure and Economic Development: Landed Property in Bengal and Theories of Capitalism in Japan*, Occasional Papers in Third-World Economic History, No. 4, (School of Oriental and African Studies, London, 1992), pp. 24–33.

3 *See* Takafusa Nakamura, *Economic Growth of Prewar Japan* (New Haven, 1983); and Ryōshin Minami, *The Economic Development of Japan: A Quantitative Study* (London, 1986).

4 The emphases on the continuity between the periods before and after industrial revolution, the trend of sustained growth before the Western impact and the survival of traditional industry after industrial revolution, are examples of this.

5 *See* Chie Nakane, *Kinship and Economic Organisation in Rural Japan*, (London, 1967), and Chie Nakane and Shinzaburō Oishi (eds), *Tokugawa Japan: The Social and Economic Antecendents of Modern Japan* (Tokyo, 1990).

6 Thomas C. Smith, *The Agrarian Origins of Modern Japan* (Stanford, 1959), Chapter 7; ditto, *Native Sources of Japanese Industrialization, 1750–1920* (Berkeley, 1988), Chapter 8. On the significance of labour absorption in Asian agriculture, *see* Shigeru Ishikawa, *Economic Development in Asian Perspective* (Tokyo, 1967), Chapter 1.

7 Susan B. Hanley and Kōzō Yamamura, *Economic and Demographic Change in Preindustrial Japan, 1600–1868* (Princeton, 1977); Thomas C. Smith, *Nakahara: Family Planning and Population in a Japanese Village, 1717–1830* (Stanford, 1977). For a more recent development, *see* Osamu Saitō, 'Infanticide, Fertility and "Population Stagnation": The State of Tokugawa Historical Demography', *Japan Forum*, 4–2, (October 1992), pp. 369–81.

8 Osamu Saitō, 'Population and the Peasant Family Economy in Proto-Industrial Japan', *Journal of Family History*, Vol. 8, spring 1983, pp. 30–54.

9 Ichirō Inukai and Aaron Tussing, 'Kōgyō Iken: Japan's Ten Year Plan', *Economic Development and Cultural Change*, 16–1 (October 1967), pp. 51–71.

10 Yōjirō Hayami and Saburō Yamada, *The Agricultural Development of Japan: A Century's Perspective*, (Tokyo, 1991), pp. 66–70.

11 James I. Nakamura, *Agricultural Production and the Economic Development of Japan, 1873–1922*, (Princeton, 1966), Chapter 7; Penelope Francks, *Japanese Economic Development: Theory and Practice* (London, 1992), p. 109.

12 Hayami and Yamada, *The Agricultural Development of Japan*, pp. 70–72; Hitoshi Saitō, *Nōgyō Mondai no Tenkai to Jichi Sonraku (The Development of the Agrarian Question and the Village Autonomy)* (Tokyo, 1989), pp. 3–48.

13 Hayami and Yamada, *The Agricultural Development of Japan*, pp. 175–97.

14 For the importance of traditional industry, *see* Takafusa Nakamura, *Economic Growth of Prewar Japan*, Chapter 3. I was unable to incorporate the recent literature on the development of the tertiary sector here except for brief references to distribution, local credit and consumer patterns, but, among Japanese scholars, there is now awareness of the importance of this topic.

15 Kanji Ishii, *Nihon Sanshigyō Bunseki (An Analysis of the Japanese Silk Industry)* (Tokyo, 1972); Ryōshin Minami and Fumio Makino, '*Seishigyō ni okeru*

Gijutsu Sentaku' (Technological Choice in the Silk-reeling Industry)', in Ryōshin Minami and Yukihiko Kiyokawa (eds), *Nihon no Kōgyōka to Gijutsu Hatten* (Tokyo, 1987) pp. 43–63. Shinya Sugiyama, *Japan's Industrialization in the World Economy, 1859–1899: Export Trade and Overseas Competition* (London, 1988), pp. 77–139.

16 Ishii, *Nihon Sanshigyō Bunseki*, p. 455.

17 Tetsurō Nakaoka, 'The Role of Domestic Technical Innovation in Foreign Technology Transfer: The Case of the Japanese Cotton Textile Industry', *Osaka City University Economic Review*, 18, (1982), pp. 45–62; Takeshi Abe, 'Traditional Industries of Japan in Early Meiji Years: The Case of the Cotton Weaving Industry', Discussion Paper 75, Faculty of Economics, Osaka University, (February 1989) pp. 9–25.

18 Yukihiko Kiyokawa, '*Nihon Menbōshi-gyō ni okeru Ringu Bōki no Saiyō o Megutte: Gijutsu Sentaku no Shiten yori*' ('On the Introduction of the Ring Frame into the Japanese Cotton Spinning Industry: A Study of Technology Choice'), *Keizai Kenkyō*, 36–3 (July 1985), pp. 214–27; Kaoru Sugihara, 'Japan as an Engine of the Asian International Economy, *c.* 1880–1936', *Japan Forum*, 2–1, (April 1989), pp. 127–45; Nakaoka, 'The Role of Domestic Technical Innovation', pp. 54–61.

19 Kaoru Sugihara, 'The Transformation of Young Country Girls: Towards a Reinterpretation of the Japanese Migrant (*Dekasegi*) Industrial Labour Force', in Janet Hunter (ed.), *Aspects of the Relationship between Agriculture and Industrialisation in Japan*, STICERD, London School of Economics, London, (1986), pp. 32–51.

20 Alexander Gerschenkron, *Economic Backwardness in Historical Perspective*, (Cambridge (Mass.), 1962); Henry Rosovsky, 'Japan's Transition to Modern Economic Growth, 1868–1885', in Henry Rosovsky (ed.), *Industrialization in Two Systems: Essays in Honor of Alexander Gerschenkron* (New York, 1966), pp. 91–139.

21 Kaoru Sugihara, 'The Development of an Informational Infrastructure in Meiji Japan', in Lisa Bud-Frierman (ed.), *Information Acumen: The Understanding and Use of Knowledge in Modern Business* (London, 1993), pp. 75–97.

22 Takafusa Nakamura, *Meiji Taishōki no Keizai (The Japanese Economy of the Meiji and Taishō Period)*, (Tokyo, 1985), p. 202. Strictly speaking, cities (*shi-bu*) and countryside (*gun-bu*) referred to here are administrative categories. There were some towns and villages, classified as countryside, which had the population of more than 20,000.

23 See Minami, *The Economic Development of Japan*, Chapter 9.

24 Kaoru Sugihara, 'Japan's Industrial Recovery, 1931–1936', in Ian Brown (ed.,), *The Economies of Africa and Asia in the Inter-war Depression* (London, 1989), pp. 152–69.

8

Agriculture and Industrialization in Colonial India

David Washbrook

The stagnation of the Indian economy during the colonial period (1800–1947) and the weakness of industrial development remain, as Paul Bairoch and Colin Simmons have remarked, a mystery that economic history has, as yet, adequately to explain.[1] From the status of one of the the early modern world's leading manufacturing and merchandising economies (with, according to Bairoch, nearly a quarter of the world's total manufacturing capacity as late as 1750), South Asia had shrunk to possessing barely 1 per cent of world industrial output and trade by the middle of the twentieth century.[2] Of course, only a small part of this decline (particularly in the second quarter of the nineteenth century) was absolute: for the most part, it represented an appearance created by the faster growth of other parts of the world economy. Colonial India's principal problem was a stagnation that kept her industrial sector unchanging at just 10 per cent of the total work-force from the 1860s to the 1960s.[3] But, as a case of relative stagnation, it has to be considered as quite spectacular.

How directly issues concerning the relationships between agriculture and industry bear on explanations of this dismal history is very much an open question. South Asia's demise was a corollary to Western Europe's rise, and most theories of India's stagnation would give pride of place to the colonial factor and to the political side of political economy. Certainly, it would require a remarkable degree of professional 'blinkeredness' to confine discussions of the the Indian case to the 'micro-level' and to abstract models of economic/market behaviour, which ignored the political context. At the same time, however, simple political explanations, based upon the supposition of mechanisms of immediate 'surplus appropriation' and 'repression', do not go very far towards clarifying the matter. The underdevelopment-theory notion that India was impoverished by means of direct surplus extraction, for example, misses the extent to which the drain of surplus (which undoubtedly took place) was marginal to the economy as a whole.[4] Statistical calculations of the colonial economy are necessarily conjectural but some recent 'guesstimates' would reckon a 35 per cent increase in per capita income between *c.*1860 and *c.*1920, mostly as the result of agricultural expansion.[5] Colonialism, at least after the early nineteenth century, did not operate on the basis of simple 'robbery'.

Equally, at least after the initial phases of conquest by the English East India Company, it is hard to find evidence of policies deliberately aimed at suppressing the development of Indian industries even in areas which were competitive with Britain. Certainly, the government was inclined to favour British firms with its contracts and licenses; certainly too, it connived at racially discriminatory practices in Indian markets; and certainly also, its ideas on 'fair' competition were tilted to guarantee British winners.[6] But, if it is supposed to have been concerned to 'stamp out' competitive industries, one must note that it was remarkably ineffective: most of the few industries that India did develop in the colonial period were in areas of immediate competition with Britain (cotton, jute) and, in some cases, were started with expatriated British capital. Further, in stamping out competitive industries, the British Indian government would also have to be credited with stamping out many non-competitive ones, which could have been developed very profitably for British capital itself. As A. K. Bagchi has argued, there were several areas of potential industrial growth in the colonial economy that remained unexploited – and unexploited even by the British![7]

The most significant sins of the Colonial State may have been of omission rather than of direct commission. Large areas of the economy were organized on almost obsessive *laisser-faire* principles, that allowed India no protection from the competition of already industrialized and newly industrializing nations (including, of course, Britain). With a single exception, State investment policies were parsimonious in the extreme: the exception, where lavishness and waste ruled the day, was the railway system, the profits of which were sustained, against the rest of the economy, by monopoly price fixing and taxation. Before at least the 1920s, the State had no policies and no interventionist mechanisms aimed at promoting industrial growth. Indeed, as we shall see, such policies would have cut across the lines of the strategy which it had developed for dealing with the 'agrarian problem'. Given the importance (at least on most scholarly views!) of the role played by State intervention in the 'second-phase' industrializations of Japan, Germany and

Russia during the late nineteenth century – and perhaps even the covert role played by it in Britain's 'first phase'[8] – it might be felt that the British Indian State's omissions, in and of themselves, provide adequate explanation of India's 'failure' without the need for reference to the specifities of the interface between agriculture and industry.[9]

And, if this 'failure' was taken to be that of not undergoing a full-scale and macro-level transformation to an industrially dominated or led economy and 'modern economic growth', then it would be hard to argue against the case. At perhaps a lesser level of significance, however, there remain some questions which consideration of the relationship between agriculture and industry might still illuminate. Given that the constraints imposed by the colonial framework did not rule out all forms of industrialization, why did it only come to take place in certain parts of India but not others? Why did the process affect certain forms of production but not others? Why were some industrial undertakings successful but others not? Why, perhaps, was there not more industrialization than there was? or, conversely, why was there ever any at all?

As soon as we begin to look at the relationship between agriculture and industry for answers to these questions, however, a second set of problems arises: just what is, or ought to be, the 'proper' relationship between the two for the promotion of economic growth? The once-celebrated 'stages-of-growth' model, for example, which surmises that industrialization follows (in some way or other) from growth in agriculture, is sharply contradicted by at least two of colonial India's economic experiences.[10] Between the 1820s and 1850s, her predominantly handicraft manufacturing sector experienced extensive 'de-industrialization', while agricultural output was increasing.[11] Conversely, in the inter-war years, India enjoyed her most rapid phase of industrialization while her agrarian economy was being severely racked, and in some cases returned to a subsistence orientation, by the forces of world depression.[12] Here, it might seem as if agriculture and industry were antagonistically related, and that the growth of one was premised on the decline of the other.

But that that provides no general rule either is suggested by the counter-experience of the years 1860–1914, when there was growth in the agricultural and in the industrial sectors of the economy and, at least with regard to jute and cotton, the two were clearly related. To complicate matters further, however, even in this period, the relationship between agricultural and industrial growth was not simple. The region in which agricultural growth was fastest (indeed, perhaps the one unequivocal 'success' story of colonial agricultural development) was Punjab, but it generated very little industry.[13] Conversely, even by the 1860s, the agrarian economy of West Bengal was in serious decline, but its local centre of Calcutta became one of India's principal industrial metropolises.[14]

As these contrasting cases suggest, relationships between agriculture and industry in India were complex, and varied greatly between time and place, making attempts to construe them outside their contexts of particularity vacuous in the extreme. Colonial 'India' was the size of Europe, and scarcely less regionally differentiated; and its history stretches across almost the whole of the 'modern' period. Yet, if historical adequacy requires particularity, the demands of this paper – and this book – call for a return to some level of generality at which the broad possibilities of the relationship might be discussed. To facilitate this latter purpose, I propose to consider the Indian case(s) in terms of the interplay between two sets of factors (those from the supply and those from the demand sides of the economy) and to use particular regional data only *inter alia* to help refine conceptualization.

The Supply Side of the Indian Economy

How well did Indian agriculture supply any notional industrializing efforts with capital, raw materials and labour? The case concerning capital is rendered difficult by several 'exogenous' variables deriving from the colonial context. In certain senses, there was no 'Indian' economy during this era but rather an Indian location within a British-imperial world economy. Capital could, and did, flow into India from other parts of this economy – expatriated British capital, as well as earnings and savings made by 'Indians' operating in other parts of the world. Agriculture was not the only source of potential capital and, hence, reference to its condition does not directly explain the availability or non-availability of resources for industrialization.

Even if these exogenous factors are left out of account, however, there appears to be little argument that Indian industrialization was hampered by any absolute shortage of agriculturally generated capital as, for example, Dwight Perkins has claimed was the case in China.[15] At no time before 1920 did India experience the levels of population growth found in Ch'ing China, which, according to Perkins, rendered virtually negative the surplus from agriculture. India's rates of population growth rarely reached above 1 per cent per annum and, throughout the nineteenth century, various forces guaranteed a continuing expansion of agricultural output.[16] In the first half of the century, these consisted principally of sectoral shifts in the work-force towards agriculture, following 'de-industrialization' and 'de-urbanisation', and the extensive clearing of forest margins for the plough; in the second half, growth was promoted by the intensification of cash cropping for world markets. Paradoxically, the one period when population growth seriously threatened the availability of a surplus from agriculture was the 1920s and

1930s – paradoxically, because these years witnessed the fastest rates of industrial growth.

India's 'problem' was less the availability from agriculture of a potentially investable surplus than the ways in which that surplus was invested, the values and goals of the groups who commanded it and the institutional structures organizing its deployment. First, and most obviously, the British Indian government took a substantial share in taxes, especially in the first half of the century. This was used, partly, to prime (via contracts and concessions) British-centred industries; and, partly, to build the framework of a world political and military empire which only marginally stimulated Indian-centred industries.[17] Second, a considerable portion of the rest was handed out to 'aristocracies' and 'royalties' (frequently created by the British for reasons of political stability) whose ethics strongly favoured consumption over investment.

Admittedly, a third and, from the middle of the nineteenth century as land revenues fell and the economic power of aristocracies waned, a growing portion passed into the hands of merchant and 'rich peasant' groups with strongly entrepreneurial leanings. Much of the industry which did come to develop, particularly textiles in west, and south India, and, in the inter-war period, jute, cement and paper industries in eastern India, was based upon their investments. There were structural problems in organizing these investments, however. The 'modern' banking system, introduced by the British, was racially shy of Indian business people and as conservative in its attitudes towards industry in India as it was in Britain. Modern Indian banks began to emerge only in the 1920s. There was, of course, a variety of Indian 'traditional' banking systems operating throughout the period, which could command substantial capital reserves and did finance industry. They tended to be caste and family specific in their clientele, however, and did not, therefore, provide means for the general mobilization of funds. Also, they had originally been built up in relation to agro-financing and transnational merchandising activities, which continued to make alternative demands on their funds.[18]

In the end, it would be hard to argue that an absolute shortage of capital, and agriculture's failure to yield a supra-subsistence surplus, accounted for the low levels of industrial growth found in colonial India. Rather, capital was being deployed, on a very considerable scale, for other purposes; and was organized in a way that made difficult the undertaking of long-term industrial investments.

Similarly, it would be hard to understand stagnation as the result of a shortage of appropriate raw materials. Outside the agricultural domain, India possessed considerable supplies of fossil fuels, although they were only marginally exploited during the colonial period. Within that domain, it came to produce vast quantities of raw materials (cotton, jute, oilseeds, tobacco, hides, woods, silk, etc.) which, exported as primary products, serviced the industries of other parts of the world. Particularly during the later nineteenth century, when the sub-continent enjoyed currency advantages, India was one of the world's major agricultural export economies, trading not only in foodstuffs but, and particularly, in a variety of industrial raw materials. Land devoted to cotton growing doubled in area and, in some provinces, trebled; jute production rose from almost nothing to the point where it covered nearly a quarter of the total cultivated land of East Bengal; groundnut production also grew from nothing until it covered a million acres of south India by the time of World War I.[19] If these raw materials could be ingested by industries centred elsewhere in the world, why, except in small quantities, could they not be used in India itself?

Admittedly, colonial export agencies occasionally noted problems of supply, which can be traced back to the conditions of a peasant agriculture lacking investment (especially in irrigation) and industrial inputs (fertilizer, etc.). Quality was often variable and difficult to control; new and improved crop strains were not easily disseminated. Over time (particularly after World War I), India came to lose a number of commodity markets through the rigidities of her systems of production which, for example, limited the supply of long-staple cottons and industrially processable sugars. Nonetheless, if the most technologically advanced systems of industrial production may have encountered problems with Indian raw materials, the less advanced systems in many parts of the world obviously did not. Notionally, there was scope for much more 'early' industrial technology in Indian manufacture than ever came to be applied. Moreover, Indian agriculture was not completely rigid: some regions, such as Punjab, continued to adapt to new industrial demands from abroad and to displace other regions less favourably endowed. Industrialization was not significantly 'arrested' by raw material problems on the supply side; rather, Indian-produced raw materials aided industrialization in practically every world economy that underwent it.

The third supply-side factor, labour, probably did present greater problems to notional industrialization efforts, although not in the form once commonly assumed. During the colonial period itself and into the 1950s, a number of economists held that Indian industrialization was inhibited by lack of adequate labour coming out of agriculture. At least before the 1930s, general urbanization levels rarely rose above average rates of population growth, and the bulk of India's 'landless labour' was retained in the countryside. Morris D. Morris readily disposed of this hypothesis, however, and demonstrated, from the case of the Bombay textile industry, that agriculture yielded an adequate labour supply, in terms of numbers and of ability to learn the requisite skills, to meet whatever demands had ever actually been made of it.[20]

More recently, however, 'problems' of a rather different kind have been traced back to questions of labour supply. The first concerns how far there may have been, not too little labour availability, but too much. Labour markets were organized only 'informally', and a common characteristic of them, at least in the two main industrial centres of Bombay and Calcutta, seems to have been a sharp division between a small sector of permanently, or long-term, employed workers and a mass of casual workers vying desperately for occasional employment. The legions of the casually employed also tended to grow over time, particularly between the wars, as agricultural depression destroyed rural employment. Wage differentials between the two groups were considerable, and due more to factors of politics and privilege in the way that labour was recruited than to those of meaningful skill. In particular, non-official 'jobbers' were used as contractors to recruit labour from their natal villages, and they built up 'private' powers and privileges in the course of their relationships with employers.

This structure of labour organization and recruitment can be seen to have hindered industrialization, or at least deterred on-going and progressive investment in industrial technology, in several ways. Cheap labour is ever the enemy of technological innovation and, faced with competitive pressures, Indian industrialists were inclined to turn to ways of lowering labour costs, particularly by breaking through the restrictive practices of jobbers to reach the pool of 'cheapest' labour beneath them. As Raj Chandavarkar has shown, the principal strategies of 'rationalization', 'scientific management' and 'improved profitability', discussed in late-colonial India in British and in Indian business circles, concerned techniques of lowering the wage bill rather than of introducing new means of production.[21] Indeed, the corollary to these strategies at the level of work practices was to make the introduction of more advanced technologies of production increasingly difficult. Wage costs were best lowered by casualizing the entire work-force: but how could a work-force, deliberately kept 'on the move' and regularly 'disemployed', ever learn advanced work and mechanical skills? Here, the propensities of the agricultural sector to make available large supplies of labour functioned to create opportunities for industrial capital to maximize its profits by investing as little as possible in industrial technology.

And labour coming from the agricultural sector affected industrial strategies in another way, too. As might be expected, Indian workers resisted strongly a logic of 'development' premised largely on their own immiseration and on the driving down of industrial wages to levels comparable with the wages to be found in a depressed rural economy. The labour relations of Indian industrialization were extremely fraught, particularly after World War I when the pace of industrialization was increasing. Here, the agricultural character of the work-force became important in that it provided resources from outside the industrial economy with which labour could defend itself. In India, labour was released from agriculture to industry, less as the result of outright 'proletarianization', than of peasant 'pauperization'. Land plots diminished to the size where they could no longer meet fully family needs, and some family members had to sell their labour. But some 'family land' was retained by vast numbers of people who ended up working in the industrial sector; and close connections with the countryside were maintained, often for several generations. These connections, together with others deriving from ethnic and community solidarities generated in 'the factory' and in the countryside, could be mobilized against the 'rationalizations' of industrial capital. Given the apparent desperate poverty of the work-force, labour resistance could be extremely strong and capable of being sustained over very long periods. Factories were often closed down for months at a time and labour 'unrest' was an endemic condition of industrial life.

As Chandavarkar again has shown, the 'power' of labour here played a significant role in determining technological innovations and changing work practices.[22] On many occasions, to keep production going, employers came to junk new machinery, which they had bought, and to accept the refusal of their labour-forces to work with it. On other occasions, they might manage to install the machinery but only at the price of agreeing to sustain unnecessary, but 'traditional', employment levels, working hours and wage differentials, the costs of which absorbed most of the supposed benefits.

Indian industry's agriculturally derived labour-force, then, may well have posed serious supply-side 'problems', although they were problems much more of a political than of an economic kind. On the one hand, agricultural depression after World War I offered industrial capitalists opportunities for the super-exploitation of cheap labour in ways that raised their profits without increasing production. On the other, the rurally entrenched resistance of labour blocked efforts to disturb the privileges and powers constructed under original and previous systems of work and recruitment. Between the two, the room left for capital to 'revolutionize' the means of production was strictly limited.

The Demand Side of the Indian Economy

In many ways it is the demand side of the economy that provides better clues to the stagnation of Indian industry: and clues, too, to the constrictions – of disorganized capital markets, rigid crop-production regimes and conflictual labour relations – found on the supply side. Put most simply, the small scale and instability of demand for industrial manufactures in Indian markets questioned the rationality, and profitability, of making industrial investments there. Of course, as part of the British imperial system, the Indian economy was not isolated, and the rationality and profitability of investments within it need to be seen also in the wider context of this system. Before World War I, the few industries that did emerge were clearly responding to export opportunities. Most of the production of the jute industry and, in its early days, of the cotton-spinning industry were for export. Such opportunities were few, however, with British finance and transport capital providing the infrastructure for most of the international economy, and offering linked service packages to British-centred industries, it required very major advantages in comparative production costs for India to offer itself as an alternative location for manufacture. Jute production in Calcutta and, for a time, cotton-yarn production in Bombay, had these advantages but no other Indian-centred manufacturing processes did.[23]

Under colonial conditions of political economy, the demand for industrial manufactures in India, itself, was extremely limited. With regard to industrially produced cotton textiles, for example, while India famously became Britain's largest single export partner, the proportion of annual British textile production taken by India before World War I never amounted to more than 20 per cent. Even had these imports notionally been substituted by local production, they would have given India the scope for a modern textile industry of only trivial size compared to that of Britain (and to the Indian economy as a whole). Much the same would seem to have been true of most other forms of industrial manufacture. Indeed, during the 1930s, after the British had belatedly allowed the Indian economy some tariff protection, this became very clear. Most of the new industries, apparently guaranteed easy profits behind tariff walls, rapidly became overcapitalized, in relation to indigenous demands for their products, and either went bankrupt or else imposed restrictive quotas on themselves.[24]

Based exclusively upon Indian domestic demand, modern industries found it difficult to make economies of scale, and were inefficient compared to their international competitors. Based exclusively on this demand, too, they faced serious problems in stabilizing production systems to meet the needs of highly variable Indian markets. As one viceroy put it, 'the whole of the Indian economy is a gamble on the monsoon', and, in years of serious drought (which came virtually every decade to one or another part of India throughout this period), demands for cloth and other industrial manufactures were the first to fall in response to rising food prices.[24] 'Imperial' industries, for which India constituted but one local market among many, could counter these instabilities by switching exports elsewhere, or by relying on sophisticated metropolitan credit systems to help them survive. But Indian-centred industries were obliged to take the full brunt without protection or safety nets. Entrepreneurs faced periodic bankruptcy as a certain condition of investment: the ownership of Bombay's cotton-textile factories, for example, changed on average once every seven years between the 1880s and the 1930s.[25] Under these circumstances, industrial investments, if worth it at all, could be conceived only as short-term gambles from which it was hoped to escape with quick profits before the bailiff inevitably called. It would have been the height of irrationality to invest heavily and continuously in plant and equipment to service market demands that vanished, literally, on the winds.

The poverty, insecurity and low productivity of India's agrarian economy undoubtedly played a major role in shaping the limited and variable schedule of demand for industrial manufactures. National income, mostly derived from agriculture, stood at about one-tenth that of Britain in the half-century leading up to 1914. While certain regional economies (particularly in the West) enjoyed a degree of progress on the basis of greater cash cropping for international markets, their gains were largely offset by crises and declines taking place elsewhere (particularly towards the East). Arthur Lewis has estimated overall growth in productivity at scarcely 1 per cent per annum even during 'the silver age' (*c.*1875–95), when imbalances in currency ratios made Indian-produced primary products exceptionally attractive on international markets.[26] Recurrent droughts and famines (for example, in south and west India in 1866–68, 1876–78, 1896–98 and 1900–01) also constantly disturbed the gradient of 'development', such as it was. If Indian industrialization waited upon the emergence of prosperity and purchasing power among the mass of its peasantry, it may be no wonder that it ended up waiting a very long time.

But has industrialization anywhere ever really waited upon the development of significant demand out of agrarian society, especially peasant society? The key factor in the histories of other industrializations would seem to be more that agricultural growth should pass expanding and secure purchasing power to élite, non-peasant, 'bourgeois' groups whose lifestyles and consumption tastes require industrially manufactured goods. The miseries and low productivities of the Russian and Prussian peasantries did not stop, indeed, may have been a condition of, industrial growth in Russia and Germany. The issue, then, may be as much one of distribution as of production and productivity. And, if so, the peculiarities of the colonial Indian class structure, which underwrote the system of distribution, come directly into focus.

The political structure of colonial India militated against the development of large aggregates of 'middling' incomes, likely to be spent on industrial manufactures and possessed of sufficient political power to insure themselves against the vagaries of the agricultural economy. The origins of the colonial structure lie, as Christopher Bayly has seen, in the break-up of

India's *ancien régime* in the second quarter of the nineteenth century. This regime had generated a huge demand for artisan manufactures and services to support the innumerable court centres and armies which competed for power within it after the fall of the Mogul Empire. These courts and armies may, indeed, have extracted a substantial surplus from the Indian peasant but they expended most of it in India itself and were principally responsible, far more than was overseas' demand, for the flourishing condition of domestic manufacture.

As British power became consolidated in the second quarter of the nineteenth century, major aims of policy included confiscating most of the revenues on which the courts were based, dismantling their armies and reducing the size of their service sectors, which were held to represent the height of wasteful, oriental luxury. Vast numbers of soldiers, servants and artisans were disemployed and pushed out into the countryside to swell the agricultural work-force. The result was a catastrophic collapse in internal demand which, probably more than the impact of the British industrial revolution and company policies of exporting Indian specie to China, was responsible for the economic depression and the de-industrialization and de-urbanization of these years.[27]

Of course, notionally, the dismantling of this 'feudal' economy ought to have given rise to a capitalist one, in which a dominant 'bourgeoisie' would lead India towards 'modernization' and industrialization. But, although there were tentative beginnings, particularly in Bengal, a strong bourgeois class formation never quite emerged, either in political or in income terms. In the first place, the dismantling of 'feudalism' proved partial as the British came to look to the supposedly 'traditional' authority of a residual and puppet aristocracy as an important bulwark to their rule. What this meant, in economic terms, was that a considerable share of the available surplus was left in the hands of maharajas and large landlords. Politically secured by the British from overthrow, however, these 'traditional' rulers no longer found it necessary to rule in traditional ways. Rather than expending their 'revenues' on building up large retinues of clients, servants and dependants in their own domains, and functioning as the centres of economic redistribution, they spent their 'incomes' on palaces and high luxuries brought from Europe to celebrate the culture they now claimed to share with their new masters. European artists, architects and artisans gained from their bounty which, in any event, was little directed towards the procurement of industrial manufactures.[28]

An Indian 'protobourgeoisie', drawn from the ranks of merchants, scribes and educated professionals, and petty rentier landlords, was left to develop on only that part of the agricultural surplus not siphoned off by the colonial rulers for their own imperial purposes or consumed by the collaborating aristocracy. This made its resource base very narrow although, for a time, the possibilities looked promising. In eastern India, the peculiarities of the Bengal Permanent Settlement left substantial extractive power over the peasantry in the hands of a *bhadralok*, or 'respectable', class/caste formation of small landlords-cum-scribal professionals oriented towards the city of Calcutta.[29] Particularly during the depression of the 1820s to 1850s, they were able to mount a ferocious 'rental offensive' against the peasantry, which passed an increasing share of the social product in their direction – a share that they noticeably spent on the purchase of industrial manufactures, albeit and necessarily, from Britain.[30] In a more limited way, too, the ratio between high rent and revenue demands and low crop prices, which marked these years, served 'protobourgeois' consuming groups and interests throughout India, and established the market for British goods.

But the ratio, and the politico-economic conditions lying behind it, were not to last. Particularly following the Great Mutiny of 1857, the British became concerned that the peasantry was being squeezed to the point of threatening agrarian revolt. A plethora of tenancy and indebtedness legislation followed (together with reductions in land-revenue demands), which weakened the force of landlord and moneylender surplus extraction from the peasantry. A greater share of potential purchasing power was left with peasant society, or at least with its upper and richer levels. But this power, held by part-subsistence producers who were culturally distant from the world of manufactures, was less likely to be directed towards industrial consumption than when possessed by the 'protobourgeosie'.[31]

A second set of forces working towards the same end can be seen in the structure of crop prices. From the 1860s to World War I, there was a continuous inflation which, in and of itself, helped to transfer surplus from 'appropriators' to 'producers' because rent, revenue and interest demands tended to lag behind real prices. Peculiarly, though, and for reasons we shall see, internal food prices (especially grain prices) rose faster than 'export' cash-crop prices and even faster than the prices of (imported) manufactured goods. This placed a progressive squeeze on the incomes of all those, such as the 'protobourgeoisie', who had to purchase their food supplies in the market-place. Just as their rental and usury incomes started to fall, so their costs of subsistence started to rise – reducing their potential consumption of industrial manufactures.[32]

Nor did they receive much compensation from the wider policies of the Colonial State. Parsimonious in the extreme, the State developed the infrastructure of modern government (education, legal institutions, welfare facilities, etc.) that provided major 'protobourgeois' employment opportunities only very slowly and partially. The 'urban' population, roughly 10 per cent of the whole, increased not a percentage point faster than the total population between 1871 and 1921. Indeed, 'protobourgeois' frustrations, particularly among the *bhadralok*, became a noticeable feature of the politics of the early twentieth century, and began to colour the Indian national movement. In the short term, however, their expression proved highly counterproductive: they led the British to identify the nationalist threat with 'protobourgeois' interests and to scheme all the harder to keep those down, favouring an India that remained 'feudal', 'peasant' and 'loyal'.

How far these distributional factors affected industrial performance before 1914, may be seen from a comparison with the 1920s and 1930s, when their balance altered again. Suddenly, economic conditions started to run strongly in favour of the 'protobourgeoisie', after three-quarters of a century of adversity. The Colonial State was finally obliged to concede the significance of the nationalist challenge and, without entirely abandoning the peasantry and the feudality, to offer policies of industrial promotion and government expansion (the latter vastly proliferating opportunities for professional employment). Equally, under the impact of the Great Depression, food prices fell faster than those of manufactured goods, and increased non-necessary consumption surpluses in family budgets (at least among those families that had surpluses).[33]

The result was an increase in 'protobourgeois' purchasing power and a small-scale explosion in domestic manufacturing industry, with Indian entrepreneurs finding it profitable to move into many of the areas of production once held as virtual British monopolies. Of course, what took place was scarcely an industrial revolution but only the beginnings of 'enclave' development for the limited, emergent Indian middle-class market. The masses were to remain excluded from most forms of industrial consumption for another generation at least. But 'enclave' development marked a distinct advance over no development at all.

The 'Problems' for the Indian Industrial and Agrarian Economies

The question of colonial India's industrial growth, then, cannot be reduced to a simple function of the failure of its agricultural economy. More was certainly always possible on the basis of what that economy yielded, little though it was. Nonetheless, the facts that productivity was so low, and available surplus so short, can be taken to define the line beyond which Indian industrialization could never have gone, even given different state policies and a distribution of wealth more conducive to growth. But why was Indian agricultural productivity so low and growth all but stagnant?

Obvious answers, focusing on demographic pressures or the 'irrationalities' of peasant production systems or the hostilities of nature, are scarcely adequate. As noted before, at roughly 1 per cent per annum before 1921, the rate of Indian population growth was slow – exceptionally so by contemporary European and Asian standards. The population increased from *c.*150 million to only *c.* 300 million between the battles of Trafalgar and the Somme. Equally, colonial rule opened out Indian agriculture to the market rationalities of the international economy and the interventions of advanced Western technology. If the Indian agrarian economy did fail, it was not for want of contact with the world of modern capitalism. Indeed, in some ways it was the depth of this contact that actually caused several of the worst problems.

The celebrated colonial construction of the Indian railway system, the second largest in the world after Russia, was a distinctly mixed blessing for the futures of industry and of agriculture. For industry, policies that raised the capital and procured the lines, rolling stock and even coal in Britain represented a great 'lost opportunity' for Indian heavy industry. The railways were also built with no intention of promoting Indian industrial growth in mind and, for that purpose, were poorly articulated. Until World War I, for example, transport costs made it cheaper for Bombay to import coal from Britain than bring it from the Bihar coalmines.[34]

For agriculture, some of the consequences of railway construction were even more difficult. Hundreds of thousands, and possibly millions, of fertile hectares were lost to careless (and cheap) engineering strategies which, for example, disturbed the drainage system of Bengal and converted once-rich ricelands into malarial swamps; and which denuded parts of north and central India of forests, thereby raising local temperature levels, increasing monsoon flooding and reducing available animal fodder. The railways also promoted the trans-subcontinental transmission of deadly diseases and played an important part in keeping down population levels.[35]

Of course, the railways did create the means for a huge expansion in export cash cropping, which brought late nineteenth-century India the greater part of its economic growth. But they brought, also, a corollary in the market for foodstuffs, which, simultaneously, reduced the likelihood that that growth would lead to industrial development, and even put lives and long-term production levels more seriously at risk. The striking escalation in food price levels was due, in no small part, to the increased transport of grain, which broke down the dearth/glut syndrome previously found in most local markets. Grain now circulated more freely to find its best price in subcontinental markets. The problem was that this 'best price' was usually caused by dearth conditions in one particular area or another. Grain flooded into the area concerned, thereby relieving its immediate condition but, by so doing, spreading dearth-level prices throughout the economy. The transport revolution brought greater stability to food prices but at no small cost.

Moreover, it was a cost that large numbers of the rural poor found impossible to meet.

At least before the 1880s, there is a case that the railways deepened the crisis of famine in interior villages. Railway-construction policies between railhead towns ran far ahead of road-construction policies, linking towns to villages, and of adequate State policies of famine relief. In famine-risk districts, there had been long-standing local strategies of grain storage against likely disaster. Such grain stores, however, were not publicly owned but held by the wealthier inhabitants for distribution to their labourers and clients. As market-drawn food prices increased, there was a noticeable tendency for these stores to be run down and for agricultural labour to become casualized. Road-transport difficulties did little to prevent this, as compensation was to be found in high prices. The railhead towns gained from the resulting flood of grain and became much better able to defend themselves against famine. In the villages themselves, however, discarded clients and casualized labourers were put at greater risk and, when famine struck, paid a heavy penalty. The 'market' was incapable of overcoming the high road-transport costs to bring grain to starvelings who, by definition, had no money to purchase it. And, in the 1860s and 1870s, at least, the Colonial State's famine policies were maliciously incompetent. The Great Famine of 1876–78 was the worst in recorded history with over a third of the population of some districts in south and west India dying in their villages.[36]

Nor was it only life that suffered in catastrophes such as these: for economic history, perhaps more importantly, it was also production. The Indian agrarian economy did not operate under 'labour-surplus' conditions and, whenever famine (or epidemic disease – that other boon of the modern transport system!) struck, production levels invariably fell by at least the levels of last population. It took the southern Deccan 30 years to recover from the effects of the Great Famine; and some localities, where deserted fields had been overrun by intractable weeds and wells had become polluted, never recovered at all. Admittedly, after the Great Famine, improvements in road construction and famine policy reduced the loss of life – and crop production – associated with drought and flood. Between 1876–78 and the Bengal famine of 1943, which took place in special circumstances, no other 'disaster' was as severe. Nonetheless, recurrent dearth conditions exercised a continual check on economic growth. And the poor at the bottom of agrarian society found themselves progressively structured out of access to the market economy and dependent, increasingly over time, on State protections and subventions to keep them alive.

In several other ways, too, the interventions of modern technology and market economy caused almost as many problems as they solved. Science, for example, was readily deployed by the Colonial State and Western business houses to improve crop varieties, qualities and mixes. But it was applied almost exclusively to export cash crops, and their production rose. Reciprocally, however, no similar effort was put into raising the production of indigenous food crops: the yields of these thereby stagnated, and went into decline around the turn of the twentieth century, and their prices rose inevitably with the steady increase in demand for them made by a steadily rising population.[37]

The benefits of irrigation development, too, were of a highly questionable character. As with the railways, careless and cheap construction policies brought about some notable disasters: for example, when canal building in north India disturbed subsurface salt strata in several areas and reduced thousands of hectares of once-cultivated land to saline marshes and deserts.[38] But, even where successfully constructed, colonial irrigation works were not necessarily of maximum economic benefit. To gain highest returns on capital at lowest costs, British engineers strongly favoured dam projects in river valleys. In south India, these greatly extended paddy production in the eastward-flowing deltas. With the opening up of the south-east Asian rice bowl, however, paddy production in south India, even with irrigation support, became simply uneconomic. By contrast, the real growth area in the southern economy was the upland interior, which depended for water on well and tank irrigation. Stubborn in its riverine obsessions, the Colonial State offered not one rupee of help to the farmers of the interior.[39] Similarly, political policies dictated that the immensely expensive canal irrigation works in west Punjab, which were best utilized for 'wet-crop' production, should be handed over to wheat-producing peasants, who were encouraged to reproduce the 'traditional' farming methods (and social relations) of the east Punjab homes whence they had come.

At least one major reason for the stagnation of the Indian agrarian economy was the way that engagement with the supposedly modernizing forces of Western technological innovation and market-rational economics led to wasted resources, ecological devastation and distorted sectoral relations. But another reason, certainly, was lack of adequate engagement with those forces; or rather a very one-sided relationship in which resources were constantly being extracted from agriculture but very little was ever put back. A variety of different mechanisms was responsible for this extraction, of which State taxation was the first in importance. As indicated previously, pre-Colonial State systems had undoubtedly raised substantial tax revenues (if never quite the 40 per cent of production often quoted in 'myths' of the Mogul Empire). As several scholars have recently shown, however, these systems were also inclined to reinvest substantially in the development of production.[40] During the crucial period from the 1820s to the 1850s, the new East India Company government sought to raise a similarly great revenue on the basis of its claims to be a successor 'Oriental Despotism' – a concept it may very well have invented to legitimize the claim. But it reinvested practically nothing in agriculture and presided over a deepening agricultural crisis in many regions. In the Madras Presidency, for example, investments in the maintenance of irrigation works, let alone their extension, came to less than one-half of 1 per cent of the taxes taken off the land.

Later in the nineteenth century, the tax burden on agriculture eased (to perhaps 10 per cent of production by the turn of the century) and substantially more was returned in investment – in irrigation and transport facilities. Nonetheless, the surplus of Indian peasant production was used far more to support a massive military machine and world political empire than ever it was to encourage agricultural growth. Indeed, as a corollary to their concept of the Oriental Despotism, the British also invented the idea of the subsistence-providing 'village community', which went on reproducing itself (and yielding a large taxable surplus) without any inputs or support from the outside world.[41]

Other institutions of the 'modern' sector related to agriculture in a similarly limited way. Marketing structures, for example, tended to pass the greater share of the notional profits of production to merchants, usurers, exporting agencies and industrial consumers than to actual commodity producers (unless the latter were also, and incidentally, any of the former). In the 1820s to 1850s, the Company had stood on its State monopoly powers to engross most of the more valuable commodity trades for itself and its British 'private-trade' partners. Later, these markets were opened to greater, but still very imperfect, competition in which most of the powers of market knowledge, credit control and opportunity cost were stacked against all but the wealthiest peasant producer.

In its way, modern industry, too, sucked the agrarian economy dry. While, as we have seen, utilizing an extensive range of raw materials produced by the Indian peasant, industry and technology reciprocally developed very little to give back to that peasant. Imperial industries, for example, pioneered no mechanical inventions to displace, or even aid, the bullock and the plough. Nor did they apply themselves to the development of well-drilling and weed-removing equipment, which could have transformed the non-riverine interior. Perhaps most centrally though, they offered nothing to offset the declines in natural soil fertility, which many aspects of the colonial agrarian system made an increasingly serious problem.

A good case can be made that Indian agricultural productivity suffered a long-term decline between the eighteenth and the early twentieth centuries, the slight upturn of the late nineteenth century notwithstanding. The processes of 'peasantization', 'de-urbanization' and 'de-industrialization' released in the second quarter of the nineteenth century, pushed a substantially greater proportion of the population on to the land and into agriculture. Their numbers were steadily swelled by demographic growth across the century. Given that agriculture under the 'old regime' had tended to be concentrated on the richest and most fertile lands available, these pressures inevitably moved more and more of it towards the margins. As these margins – of forest and pasture land – were cleared, further difficulties emerged. Forest and pasture lands had been used for the breeding and maintenance of livestock. In many areas, as they were cut back and put under the plough, livestock ratios and animal inputs into cultivation fell. Forests also had been valuable for providing various kinds of 'green' manure and natural fertilizer, the use of which came noticeably to decline. Moreover, some of the fastest-expanding cash crops of the colonial period were notoriously soil exhausting. Unless grown in soil-replenishing mixes, for example, cotton could be ecologically devastating. In the central Deccan, the over extension of cotton cultivation in response to the high prices of the American Civil War years, left the land permanently damaged.[42]

Admittedly, there was a number of countervailing forces that in some regions, held back decline or even raised soil fertility levels for a while. In some areas, 'virginal' margins consisted of nutrient-rich land that could offer a temporary boost to productivity – until the nutrients were used up. Certain new cash crops, such as groundnuts, were soil replenishing and, if interchanged with cotton, could sustain fertility without the increasingly uneconomic practice of leaving fallows. Irrigation, too, where applied successfully, transformed the soil. In the absence, however, of the chemical intervention which Western science and industry failed, at this time, to offer the Indian peasant, the evidence would seem to suggest that peasant farming in a great many regions faced a constant uphill battle to maintain yields, particularly of food crops which rarely covered less than 70 per cent of cultivated land. And it was a battle that was lost in more cases than it was won.[43]

Drained of resources by the 'modern' Colonial State and world economy, and receiving very little in return, the development of Indian agriculture depended very much upon the initiative, entrepreneurship and activities of the 'peasants' directly engaged in it. And it needs to be said that some of them, at least, made an extremely good job of it: conjuring economic growth out of a context marked by deeply adverse circumstances. In certain agricultural zones, especially towards the west (Punjab, Gujerat, interior Tamilnadu), the period from the 1860s to 1914 witnessed a significant rise in productivity based upon the expansion of cash cropping, the introduction of new crops and the investment of 'private' capital (particularly in wells and water courses, cart transport, and crop-processing equipment such as cotton gins and groundnut decorticating machines). Admittedly, these advances came only on the lands of the 'upper' or wealthier peasantry; on those of people who perchance could obtain (from merchants and moneylenders) long-term credit on easy terms. But the activities of this small stratum did have growth implications for the rest of these local economies and, at least in Gujerat and interior Tamilnadu, started to promote small-scale industrial development as well. By the early twentieth century, agriculturally derived profits were being invested in local cotton-spinning factories and in extended machinery-maintaining workshops across these regions.[44]

In national terms, however, the progress of these areas tended to be swallowed up by the stagnation, and even 'regress', of other areas, especially towards the East (Bengal, Bihar, East Uttar Pradesh, East Tamilnadu). Here, although there was also expanded cash cropping (jute in east Bengal, sugar in Bihar), its productivity implications were minimal, and the development logic which it provided was one of 'involution' rather than of growth, with peasant society seeming to work harder and harder on less and less and poorer and poorer land just to keep its head above water – in the case of the Bengal deltas, quite literally![45] Moreover, over time but particularly after World War I, these 'involutary' tendencies started to creep westwards, too, and to threaten, in the 1920s and 1930s, many of the gains apparently made in earlier days in the west.

Behind the contrasting nineteenth-century histories of (very roughly) east and west, it is possible to see two sets of forces at work. The first consisted of a complex of ecological and demographic inheritances, which saw the west enter the nineteenth century with much lower population densities and much higher 'reserves' of nutrient-rich virgin land and well-accessible subsurface water. It was, preponderantly, towards the east that the great Indian rivers flowed;[46] and it was in their basins, valleys and deltas that 'classical' Indian civilization had always made its home. The society of the plains, forests and hills was historically much less dense and its economic resources were far less exploited, leaving room to grow and expand across the nineteenth century – at least until population pressure and the exhaustion of virgin land and well-accessible water began to catch up with initial trajectories of development.

The second set of factors concerned the proclivities of their respective social structures, received from history but strongly accentuated by the early political policies of colonial rule. In the east, landlord-gentry forms of dominance over the peasantry, based upon rental systems of exploitation, had long emerged and were quickly absorbed into, and developed by, the Colonial State. The Bengal 'rent offensive' of the 1820s to 1850s, noted above, was premised upon the exploitation of this structure, which permitted population increase among the peasantry to be rapidly converted into the charging of competition rents and heightened rates of surplus appropriation. Later in the century, in reaction to fierce peasant resistance, this structure began to give way, but to one which was not more noticeably egalitarian nor left many more resources in peasant hands. Rather, as commercial cash cropping spread and the need of the peasantry for production credit increased, usury capital inserted itself into the old structure and turned erstwhile dependent tenants into dependent consumption debtors. The great bulk of agricultural production in the east took place on the smallholdings of peasant farmers who lived and worked off the credit that was advanced to them by a variety of agencies against virtually the whole of their ultimate harvest. The peasantry here had few means beyond their labour to invest in production; and, while population growth kept production increasing, landlords and usury capitalists had no interest, and saw no need, to make costly and risky investments themselves.

In the west, landlord-gentry class structures were much less developed, and effective control over the land lay with peasant clan systems. These went some way towards deterring the 'revenue offensive' which the Company State launched against them in the 1820s to 1850s, and hence were able to hold on to more of their surplus. Of course, clan structures were not perfectly egalitarian but had their own internal hierarchies and relations of dominance over non-clan members. These were strengthened over the course of the nineteenth-century, producing a distinctive 'upper tier' of peasant families in most villages. It was members of this upper tier who were best placed to take advantage of the commercial opportunities that opened up after the 1860s. With the resources to command better market prices and cheaper credit, they ploughed their surpluses back into investments in wells, cattle, carts, cash crops and crop-processing machinery – at least while prices were buoyant and fertile land remained available.[47]

There were in-built tendencies, however, for them to develop towards the 'involutary' practices of the east. As population levels increased and the mass of pauper peasants came to need production credit, so opportunities appeared for richer farmers to deploy their surpluses as usury capital: from the late nineteenth century, the 'rich farmer' took over from the professional moneylender-merchant as the principal source of credit in these regions. Demographic pressures also increased the possibility of letting out land on tenancies rather than working it directly, which served 'upper peasant' status ambitions because most Indian élites followed a very strict 'non-manual labour' ethic. When crop prices collapsed, particularly after World War I, and commercial agriculture lost its profitability, these tendencies became dominant, and what had once seemed a highly significant transformation of agriculture came juddering to a halt. By the 1930s, west and east were moving along convergent lines, and Indian agriculture in general had reached the acme of its colonial development in a deepening crisis of production[48].

Conclusion

In 1790, James Grant stood at the apex of the Godavari delta and considered, with pride, that the English East India Company and the nascent British Empire had acquired 'a country' of such richness and agricultural fertility that only the Nile Delta compared with it.[49] Just 140 years later, his colonial successors looked down, from the same vantage point, at an agrarian slum; piled high with unsaleable crops produced at lower rates of productivity by a population three times as large – and seething with social unrest. What had gone wrong?

At one level, the 'story' of Indian agricultural development under colonial rule is one of arrogant technological mistake, brutal surplus appropriation and studied neglect of the conditions of peasant production and subsistence. The 'blame' lies firmly on the Colonial State and the various class interests that it served – British and Indian. But at another level, however, it has to be asked whether the story is one of failure rather than of success, and very considerable success, for the logic of capitalist development and the 'appropriate' relationship which industrial capital would like to establish with agriculture. In spite of its stagnant production systems, for example, there can be no doubt that a continuing and growing surplus came to be extracted from Indian agriculture – by the Colonial State, metropolitan business and industrial interests, and the Indian collaborative-ruling classes – throughout the nineteenth century and at least until the time of World War I. Thereafter, the demise of agriculture, particularly the decline in its product prices, opened the way for a rapid rate of industrial expansion. While Indian agriculture may have 'failed' the mass of the Indian population, and especially the peasantry, the nature of its development succeeded for the British Empire and for metropolitan and indigenous Indian capital.

Behind these curious successes, and representing perhaps the quintessence of the logic of capitalist development in agrarian India, it is possible to see a process that raised the profitability of capital in agriculture, not by increasing the size and value of its product, but, rather, by constantly cheapening its costs of production. In particular, the share of labour (including peasant labour) in the social product declined steadily and continuously throughout the period – and speeded up dramatically in the inter-war years to the point where it made possible industrial expansion.

Between the 1820s and the 1850s, this process of 'cheapening' was accomplished through the most obvious and brutal methods conceivable: de-urbanization and de-industrialization in the context of chronic price depression. Millions of erstwhile non-agricultural workers were pushed out on to the land to produce crops, the profits from which were scooped off by mechanisms of high taxation, international currency manipulation and monopoly. Later, with the growth of market competition from the 1860s, labour's share of the social product was kept depressed by laws of 'private property', which reduced its access to the means of subsistence, and by demographic pressures, which intensified the struggle for work and land. There is little evidence of any increase in real agricultural wages between 1860 and 1920 in spite of a *c.* 35 per cent rise in notional per capita national income.[50]

In a context of peasant petty commodity production, however, it may be less the history of wages than the history of non-wages, that provides the better clues to the logic of 'development'. Most land was cultivated by peasant families who sought from it, first of all, their own subsistence needs. Capital took these needs, and the internal power structures of the peasant family, as its major point of entry into agricultural production, which it exploited principally through them. As per capita land availability shrank, peasant families found themselves obliged to meet their needs off smaller and smaller plots. Very often, the only way that they could do this was by concentrating their efforts on the higher-value market crops which brought greater returns per unit area. But such crops were expensive to grow and highly at risk from the vagaries of international markets. The peasantry was obliged to engage usury capital in its search for subsistence and, as a result, was drawn into structures of consumption debt whereby crops were hypothecated in advance to the usurers who took whatever profits there were for themselves.[51]

Of course, these relations put usury capital itself at risk, for it now bore the brunt of market and of climatic instabilities. The benefit, however, lay in the means by which this capital gained access to the reserves of 'unpaid' labour inside the peasant family. A noticeable feature of many of the new and expanding cash crops of the colonial era was the way in which they made demands for female and child labour: cotton, jute, (transplanted) rice and certain kinds of oilseeds were notoriously importuning of quick fingers and bendable backs. Female and child labour (much of it previously employed in industrial by-employments such as spinning) became much more important to Indian agricultural production than ever before. And what made this particularly advantageous to capital was that it was paid less than male labour, if, indeed, it was paid at all. Its own subsistence and reproduction costs were given inside those of the peasant family. By attaching itself to this family, and redeploying its labour for agricultural production, capital gained a huge bonanza of effectively unpaid labour.[52]

In the 1950s, agricultural economists began to 'unlock' the secrets of colonial India's logic of development when they discovered the celebrated 'inverse farm size:acre productivity ratio'. The smaller farms became, the harder was their land worked to meet their owners' subsistence needs. As farm sizes had shrunk continuously across the colonial era, this may be seen to have generated a rising volume of per unit area production. As soon as profitability factors were considered, however, it became clear that the farm size: profitability ratio in India was not inverse but direct. High-productivity small farms were unprofitable – if labour costs were imputed to unpaid family labour, quite massively so.[53] What this, in its turn, can be seen to have meant is that the markets that consumed small peasant production (and small farms tended to be more market oriented than large ones) obtained it 'cheaply', at considerably less than its real costs of production. In effect, through their relationships with Indian peasant farming, metropolitan and industrial capital can be seen to have established, from their point of view, the ideal nexus with agriculture: a nexus that provided them with expanding volumes of production at ever declining levels of real price – at least until those levels had sunk so low that they destroyed the Indian agrarian base, whereupon this capital moved on to find pastures new.

Notes

1 P. Bairoch, 'International Industrialization Levels from 1750 to 1980', *European Journal of Economic History*, 11:2, (1982); C. Simmons, 'Industrialization and the Indian Economy', *Modern Asian Studies*, 19:3, (1985).
2 P. Bairoch, 'International Industrialization'.
3 J. Krishnamurthi, 'The Distribution of the Indian Working Force' in K. N. Chaudhuri and C. Dewey (eds), *Economy and Society* (New Delhi, 1979).
4 K. N. Chaudhuri, 'India's International Economy in the 19th Century', *Modern Asian Studies*, 2:1, (1968).
5 A. Heston, 'National Income' in D. Kumar (ed.), *Cambridge Economic History of India*, II, (Cambridge, 1983).
6 A. K. Bagchi, *Private Investment in India 1900–1939* (Cambridge, 1972).
7 Bagchi, *Private Investment.*
8 Eg. *see* A. Hopkins and P. Cain, 'Gentlemanly Capitalism and British Expansion Overseas', *Economic History Review*, 39:4, (1986).
9 *See* A. Gerschenkron, 'Economic Backwardness in Historical Perspective' in B. Hoselitz (ed.), *The Progress of Underdeveloped Areas* (Chicago, 1952).
10 W. Rostow, 'The Take-off into Sustained Growth', *Economic Journal*, 66, (1956).
11 C. Bayly, *Indian Society and the Making of the British Empire* (Cambridge, 1989), Ch. 4.
12 B. R. Tomlinson, *The Political Economy of the Raj* (London, 1979); C. J. Baker, *An Indian Rural Economy 1880–1955* (Oxford, 1984); Bagchi, *Private Investment.*

13 T. Kessinger, 'North India' in Kumar, *Cambridge History*, II.
14 R. Ray, 'The Crisis of Bengal Agriculture 1870–1927', *Indian Economic and Social History Review*, 10:3, (1973).
15 D. Perkins (ed.), *China's Modern Economy in Historical Perspective* (Stanford, 1975).
16 L. Visaria and P. Visaria, 'Population 1757–1947' in Kumar, *Cambridge History*, II.
17 D. Kumar, 'The Fiscal System' in Kumar, *Cambridge History*, II.
18 *See* Baker, *Rural Economy*, Chs 4, 5; Tomlinson, *Political Economy*, Chs 1,2.
19 For regional histories, *see* Baker, *Rural Economy*; S. Guha, *The Agrarian Economy of the Bombay Deccan* (New Delhi, 1985); S. Bose, *Agrarian Bengal* (Cambridge, 1986); E. Stokes, *The Peasant and the Raj*, (Cambridge, 1979); J. Breman, *Of Peasants, Migrants and Paupers* (New Delhi, 1985).
20 M. Morris, *The Emergence of an Industrial Labour Force in India* (California, 1964).
21 R. Chandavarkar, *The Origins of Industrial Capitalism in India.* (Cambridge, 1994).
22 Chandavarkar, *The Origins of Industrial Capitalism.*
23 M. Vicziany, 'Bombay Merchants and Structural Changes in the Export Community 1850–1880' in Chaudhuri and Dewey, *Economy and Society*; *also*, Tomlinson, *Political Economy*, Ch. 3.
24 Tomlinson, *Political Economy*, Ch. 1.
25 Chandavarkar, *The Origins of Industrial Capitalism.*
26 W. A. Lewis, *Tropical Development* (London, 1970).
27 Bayly, *Indian Society*, Ch. 4.
28 *See* R. Jeffrey (ed.), *People, Princes and Paramount Power* (Oxford, 1978).
29 B. Kling, 'Economic Foundations of the Bengal Renaissance' in R. van M. Baumer (ed.), *Aspects of Bengali History and Society* (Hawaii, 1975).
30 Bose, *Agrarian Bengal*, Ch. 2.
31 *See* my 'Law, State and Agrarian Society in Colonial India', *Modern Asian Studies*, 15:3, (1981).
32 M. McAlpin, 'Price Movements' in Kumar, *Cambridge History*, II.
33 Baker, *Rural Economy*, Ch: 3.
34 Bagchi, *Private Investment*, Ch. 1.
35 *See* Bose, *Agrarian Bengal*, Ch. 1.
36 For details (although not, in this author's view, an adequate assessment) *see* M. McAlpin, 'Railroad, Cultivation Patterns and Foodgrain Availability', *Indian Economic and Social History Review*, 12:1, (1975); *also*, my 'Economic Stratification in Rural Madras' in A. Hopkins and C. Dewey (eds), *The Imperial Impact* (London, 1978).
37 G. Blyn, *Agricultural Trends in India 1891–1947* (Philadelphia, 1966).
38 E. Whitcombe, *Agrarian Conditions in North India* (California, 1972).

39 Baker, *Rural Economy*, Chs 3, 6.

40 *See* C. Bayly, *Rulers, Townsmen and Bazaars* (Cambridge, 1983); A. Wink, *Land and Sovereignty in India*, (Cambridge, 1986).

41 On the ideological history of the village community, *see* C. Dewey, 'Images of the Village Community', *Modern Asian Studies*, 6:3, (1972).

42 Guha, *Agrarian Economy*; Baker, *Rural Economy*; Bose, *Agrarian Bengal.*

43 Baker, *Rural Economy*, Ch. 3.

44 Baker, *Rural Economy*, Ch. 3; Breman, *Of Migrants.*

45 Ray, 'Crisis'; Bose, *Agrarian Bengal*, Chs. 4, 5.

46 The exception to this is the Indus valley.

47 The division is suggested in Stokes, *The Peasant and the Raj*; and amplified for south India in Baker, *Rural Economy.*

48 Baker, *Rural Economy*, Ch. 5.

49 *See* extracts from J. Grant's report on Northern Circars in W. Firminger (ed.), *Fifth Report on the Territories of the East India Company* (Calcutta, 1918).

50 D. Kumar, *Land and Caste in South India* (Cambridge, 1965); Guha, *Agrarian Economy*, Ch. 5.

51 Bose, *Agrarian Bengal*, Chs 4, 5.

52 M. Atchi Reddy, 'Female Agricultural Labour in Nellore', *Indian Economic and Social History Review*, 20:1, (1983).

53 *See* K. Bharadwaj, *Production Conditions in Indian Agriculture* (Cambridge, 1974).

9

Soviet Agriculture and Industrialization

Mark Harrison

The Resource Contribution of Agriculture

The idea of industrialization supported by a government transfer of resources from agriculture owes much to Russian and Soviet history. In the nineteenth-century, Imperial government officials stressed the role of agriculture in supplying food for the urban population, taxes to pay for government support of the industrial sector and exports to pay for industrial technology from abroad. Populist critics stressed the extent to which government was buying industrial modernization at the expense of peasant sacrifice and agricultural stagnation.

After the Russian Revolution, in the inter-war years, Preobrazhensky (Trotsky's economic adviser), then Stalin himself stressed in different ways the possibility of paying for public-sector industrial investment programmes out of peasant incomes. Preobrazhensky's views were formed in the mid-1920s in the context of a mixed economy; he considered that an agricultural surplus could be generated for public investment by means of taxation of farm incomes and non-equivalent exchange (pushing up the prices of manufactures on the rural-urban market to make the peasants buy dear and sell food cheap).

Stalin, at first opposed to this idea, came round to the same general orientation in 1928–29. The context was now one of headlong transition from a mixed economy to a system dominated by public and co-operative ownership, increasingly regulated by physical controls. Instead of taxation and non-equivalent exchange through the market, Stalinist methods of getting resources out of agriculture relied more on simple confiscation of food surpluses.

In the post-World War II era, Western historians (in particular, Alexander Gerschenkron and W. W. Rostow) placed much stress on the idea of industrialization supported by a government transfer of resources from agriculture as a basic continuity in Russian history from the Tsars to the Bolsheviks.[1] But the reality is that this idea was never applied successfully in Russia, and was never shown to work. Neither before nor after the Revolution has there been demonstrated any direct link from forced saving of the peasantry to industrial capital formation.

Before the Revolution, the expansion of Russian agriculture was not unduly retarded by the pressure of taxation. The peasantry of the 1880s and 1890s was not, on the whole, impoverished by heavy taxation,[2] although not all regions and branches prospered. After 1885, the growth of agricultural output and of village consumption of foodstuffs was substantially faster than that of total population (and, still more, of the rural population).[3]

As for the supposed budgetary mechanism for transferring resources from agriculture to industrial investment, the rhetoric exceeded reality. The bulk of government revenues was raised by taxing the urban retail market, not peasant incomes and assets.[4] In terms of government expenditure, the greater part went to pay for military and bureaucratic items; at least 90 per cent of non-defence capital formation was normally carried out by the private sector. If there was a distinctive feature, by European Standards of Imperial budgets of the pre-World War I era, it was simply the large share of national income that government consumed, not the contribution that it made to industrialization.[5]

By the 1920s, a Soviet government had come to power which differed from pre-Revolutionary Imperial governments in a multitude of ways. Among the latter was its readiness to commit a really significant share of budget revenues to paying for industrial capital formation. For example, the 1924–25 USSR State budget allocated nearly one-fifth of total outlays to 'finance of the national economy' (mainly public-sector industry, transport and construction) compared with little more than one-twentieth of total outlays under equivalent headings in the 1913 Imperial budget. This represented an increase of spending on the economic infrastructure in real terms of about 40 per cent.[6]

The problem of the 1920s was not a lack of governmental will to divert agricultural resources into industrial modernization, but a supply-side constraint. From the point of view of the regime's new priorities, agricultural resources seemed more inaccessible than ever. While output tended to recover from the post-war famine of 1922, the sale of food surpluses to the urban market did not. Much more than before the Revolution, food surpluses were retained within the village.

There are several possible candidates for an explanation of this difficulty.[7] The Revolutionary destruction of large-scale 'commercial' farming may have been a factor. Peasant control over land and food surpluses certainly benefited from a reduced burden of rents and taxes. Agriculture's terms of trade with industry had worsened; this damaged peasant incentives to make food surpluses available to the domestic market, but did not enhance industrial profits because peasant losses were swallowed up in increased industrial costs.

Moreover, as the 1920s proceeded, the difficulty of getting food intensified because market equilibrium was increasingly disrupted by the rapidly growing volume of public-sector investment. The economic system was changing, away from a market mechanism regulated by money, prices and taxes, to a 'shortage' economy subject to non-price regulators and quantitative targets. In agriculture, this process meant a transition to direct controls, first over food surpluses, then over the food producers themselves.

Collectivization and its Results

Matters came to a head in July 1928, with Stalin's decision to secure a 'temporary tribute' from agriculture. This had three results for policy, often conflated under the general heading of 'collectivization', but best considered separately and taken in order.[8]

First was the move to 'a new procurement system' for obtaining rural food surpluses. There was an impromptu resort to coercion in the Ural region and in western Siberia in the spring of 1928; then coercive methods were extended to the country as a whole, and codified in the criminal law in June 1929. A nationwide system of compulsory food procurements was instituted which, at first, produced a great increase in peasants' food deliveries, then went too far; a crisis of rural subsistence was induced. The countryside was stripped of food and of animal feedstuffs. The nutritional standards enjoyed by the village population shrank to the level of basic physiological maintenance and below. The fodder shortage resulted in forced killing of livestock on a huge scale; the loss of animal tractive power resulted in a growing inability to carry out tasks of grain cultivation on time. Supply-side disruption culminated in harvest failure in 1932. Harvest failure combined with the forcing of food deliveries to produce famine in the Ukraine and north Caucasus. The killing of livestock in Kazakhstan meant loss of the main source of food for the formerly nomadic population. Differing assumptions about birth-rates in the famine years, 1932–33, give rise even today to widely divergent estimates of the resulting number of famine deaths – as many as 8 million if fertility was maintained, so that many babies were born only to die within a short period from hunger; or no more than 4 to 5 million if fewer babies were born and died.[9]

The second element of Stalin's turn was 'the liquidation of the kulaks as a class'. The property of kulaks (the more prosperous stratum of petty capitalist farmers) was to be confiscated, and the kulaks themselves socially isolated and excluded from the new rural order. A decree of February 1930 divided the kulaks into three grades, respectively subject to exclusion from the village, deportation into the remote interior, and, in the most 'socially harmful' cases, confinement in forced labour camps. Eventually, 381,000 families (up to 2 million people) were processed in the second and third categories.[10]

This campaign was designed to break resistance to new controls over village life and the rural economy. The kulak was the traditional leader of village opinion, the social and economic model of individualistic self-betterment to which ordinary peasants aspired. The attack on the kulaks was a message to the others. In the past, the peasant who wanted to do well tried to get on as an individual, to rise to the status of a kulak. Now the route of individualistic, competitive self-betterment was closed off for ever. From now on, the peasant who wanted to get on under Soviet power would prosper, if at all, only as a member of the collective, on Soviet terms.

The third element of new policy was 'collectivization' itself. The first Five Year Plan, adopted in April 1929, incorporated relatively modest targets for collectivization; by 1932–33, collective farms were to include 18 to 20 per cent of peasant households and some 15 per cent of the sown area. This degree of collectivization was to be achieved on the basis of advances in farm mechanization and electrification. But what happened far exceeded the plans. Within months, a relentless upward pressure, from the Stalinist leadership above and from local officials below, drove the targets higher and higher. In December 1929, a drive for all-out collectivization was launched; within three months, more than half the peasant farms in the country had been incorporated into collective farms. This great leap into the unknown brought chaos and disorder in its wake, and was followed, in March 1930, by a temporary retreat; then, in the autumn of 1930, the campaign was resumed. This time there was no further let-up. By mid-1931, the high water mark of March 1930, had been regained, and, thereafter, the percentage of collectivization rose steadily year by year until, by 1936, only one-tenth of households and a still smaller fraction of sowings remained outside the public and collective farm sector.

The collectivization process in the widest sense directly changed the way of life of 120 million villagers, and powerfully affected the role of the agrarian sector in the Soviet economic and political system. Four main effects may be distinguished. First are the effects on production, which can be seen in Table 9.1.[11] The arable sector suffered a disaster. As far as grain cultivation was concerned, a major negative factor was the loss of animal tractive power. Here, the transition to a new system of food procurements dealt a double blow: it not only took away grain from human consumption, but also, by stripping the countryside of animal feedstuffs, undermined the arable sector on the supply side.

Second, collectivization converted agriculture and the peasant into residual claimants of food. In the 1920s, the peasants met their own needs first, while the towns and the export market had to make do with what was left. In 1927–28 and 1928–29, after deduction of more than 10 million tons of centralized and decentralized grain collections, more than 50 million tons of grain remained at the disposal of the peasants. In 1931–32, the rate of collections reached nearly 23 million tons, more than double the rate of the late 1920s, but, with the decline of the harvest, the peasants' residual collapsed to only 33 million tons.[12] By the end of the year, there were famine conditions in the Ukraine and north Caucasus. Industrial workers, however,

Table 9.1 Soviet arable products and livestock, before World War I (within inter-war frontiers) and 1928–40

(a) *Arable products (million tons)*

	Grains	*Potatoes*	*Vegetables*	*Sunflower seeds*	*Sugar beets*	*Cotton fibres*	*Flax fibres*
1909–13 average	68	–	–	–	9.7	.68	.26
1913	79	29.9	8.6	.74	10.9	.74	.33
1928	63	45.2	10.5	2.13	10.1	.79	.32
1929	62	45.1	10.6	1.76	6.2	.86	.36
1930	65±3%	44.6	13.9	1.63	14.0	1.11	.44
1931	56±9%	40.6	16.8	2.51	12.1	1.29	.55
1932	56±10%	37.2	17.6	1.13	6.6	1.27	.50
1933	65±4%	41.3	17.4	1.14	9.0	1.32	.36
1934	68	43.8	17.6	1.15	9.9	1.20	.37
1935	75	60.5	12.4	1.22	16.0	1.77	.40
1936	56	44.4	8.2	1.12	16.4	2.47	.33
1937	97	58.7	15.4	1.75	21.6	2.58	.36
1938	74	38.3	6.8	1.61	16.2	2.63	.31
1939	73	40.7	9.7	2.07	14.3	2.70	.38
1940	87	64.7	–	2.41	16.9	2.19	.27

(b) *Livestock (millions, 1 January)*

	Horses	*Cattle*	*Sheep, goats*	*Pigs*
1914	37.0	55.6	90.3	19.8
1928	32.1	60.1	107.0	22.0
1929	32.6	58.2	107.1	19.4
1930	31.0	50.6	93.3	14.2
1931	27.0	42.5	68.1	11.7
1932	21.7	38.3	47.6	10.9
1933	17.3	33.5	37.3	9.9
1934	15.4	33.5	36.5	11.5
1935	14.9	38.9	40.8	17.1
1936	15.5	46.0	49.9	25.9
1937	15.9	47.5	53.8	20.0
1938	16.2	50.9	66.6	25.7
1939	17.2	53.5	80.9	25.2
1940	17.7	47.8	76.7	22.8

Source: R. W. Davies, M. Harrison, S. G. Wheatcroft (eds), *The economic transformation of the USSR, 1913–1945* (Cambridge, 1994), 286–9. Figures are estimates recently revised by Wheatcroft (grain and potatoes) or USSR Goskomstat (other products and livestock).

were assured of bread and potatoes, and industrial expansion proceeded on this basis.

Third, despite Stalin's aspiration to the contrary, collectivization failed to increase the 'tribute' from agriculture. This failure was unexpected, and can be ascribed to the manifold leaks in the new control system. On the one hand, peasants maintained access to the 'second economy' of unregulated market transactions; most peasants had become worse off but the few that still had food surpluses to sell could command very high scarcity prices, so that the terms of trade were not, after all, turned against the peasant. On the other hand, the diversion of livestock to slaughter necessitated an increased State supply of machinery services to agriculture – resources which otherwise would have been available for purposes of industrialization; the same applied to public food stocks which were belatedly returned to the countryside as famine relief.

Fourth, through collectivization, the Soviet State learned to do something that previous generations of Bolsheviks had argued was impossible and would precipitate an overthrow of the regime: to impose its will on about 120 million peasants. A price was paid for this in the hyperactivity of the security organs and uncontrolled expansion of forced labour camps, beginning in 1930, to cope with the inrush of peasant detainees. The peasantry as a whole became alienated from the Soviet system, especially in the Ukraine.

The Kolkhoz – Model and Reality

The *kolkhoz* (collective farm) was the new institutional form for control of food surpluses – but what form should the *kolkhoz* take? There was no blueprint or working model of a *kolkhoz* to guide the collectivization process which, instead, was led by a kind of radical Utopianism. This was expressed through a number of issues.[13]

How large should a *kolkhoz* be? Some activists advocated large-scale farming as an attempt to eradicate village boundaries and turn the peasant outwards from traditional, parochial horizons to involvement in society as a whole. This current was expressed in a trend towards super-large multi-village farms of tens of thousands of hectares (compared to the typical precollective farm of 25 hectares or so).

How far should socialization of property extend? All collective farms absorbed productive assets – land, basic implements and livestock. Some attempted to eradicate the family itself, as the traditional basis of private property, by transferring even family household goods and family functions to refectories, dormitories and crèches.

How should rewards be distributed? Cases were commonly reported of attempts to eradicate individualist striving altogether through an egalitarian policy of distribution of farm income to members only according to need, not according to contribution.

How far should peasant economic activity be controlled from above? Here there was an early proliferation of controls, attempting to block off all channels of individual initiative not directed through the collective. Thus, household plots and commercial activity were prohibited, and the rural artisan sector was destroyed.

In each of these respects, the initial impetus of radicalism went too far, and provided occasion for subsequent retreat in the years after 1930. Thus, as far as scale of organization was concerned, the village-level *kolkhoz* became the norm, though supplemented by multi-village organization of machinery services, grain collections and political control through the public-sector Machine Tractor Station (MTS). Egalitarian distribution was supplanted by a work-point system which entitled workers to a dividend share in the residual net output of the farm, providing at least a weak relationship between effort and reward. In 1932, the right to a family allotment was revived, and family members were also conceded the right to sell privately produced food surpluses at high scarcity prices on the unregulated '*kolkhoz* market'.

Such nods in the direction of moderation came too late, however, to stave off the worst results of collectivization. For the remaining inter-war years, the *kolkhoz* system was held together mainly by coercion. The asset losses of 1929–30 could not be made up by more pragmatism in the formation of agrarian institutions, or by an improved incentive structure. In the early 1930s, the rapidly expanding towns and industrial work-force had to be fed from a totally inadequate supply of food. The famine year of 1932 was marked by a return to harsh repression in the countryside, including extension of the death penalty to acts of theft against collective farm property such as gleaning in *kolkhoz* fields.

After 1932, there was a recovery but the progress recorded in the years 1933–37 was not sufficient to restore the situation. This meant that agrarian policy presented the regime with a continuing dilemma. The dilemma was clearly expressed in 1939 when, on the one hand, new laws were framed to compel all peasants to work at least a compulsory minimum of work-points on the collective farm. The strengthening of coercion from above was matched, on the other hand, by simultaneous initiation of a public debate on the possible decentralization of collective-farm management and rewards to the small, family-sized production unit (the *zveno* or 'link').[14]

Agriculture Becomes More Like the Economy as a Whole

From the 1930s through World War II and the early post-war period, there was no stabilization of the *kolkhoz* environment. In wartime, official stress on compulsory labour and procurements was offset by an opposing tendency of the private sector to encroach on the collective sphere. Soon after the war, in 1947, Stalinist policies sought to stiffen the *kolkhoz* regime again, and repress once more the private sector, without improving economic returns to peasant labour from the collective sector. A brief renewal of the pre-war flirtation with the idea of farm management decentralized down to the family-sized unit (*zveno*) in 1950 was firmly squashed by orthodoxy. In addition, agricultural policy was plagued by the false science of Michurinist plant biology promoted by Lysenko.

The brutal and unimaginative policies associated with Stalin would change with the latter's death in 1953 and with the efforts of his successors to relax the rigours of the Stalinist dictatorship, modernize the economy and win popularity. Agricultural policy remained unstable and went through many fluctuations of detail; and, for a time, Lysenkoism found renewed favour. Nonetheless, there emerged a basic continuity from the Khrushchev period

(1956–64) through the Brezhnev period (1964–82). Rising priority was attached to improving the quality and variety of Soviet diet and food supplies; this resulted, in turn, in the conversion of agriculture from a low-to a high-priority activity. Basic themes of policy comprised maintaining the basic system while improving the economic security of the rural population and reversing the flow of resources out of agriculture.[15]

After Stalin, the collective farm remained a basic unit of agricultural organization, but, nonetheless, the agrarian sector experienced important reorganizations. Table 9.2 shows that the private sector declined steadily in

Table 9.2 The structure of Soviet agriculture: selected years, 1928–85 (per cent of total)

	1928	*1940*	*1950*	*1970*	*1985*
Kolkhozy *and inter-farm enterprises*					
Sown area	1	78	83	48	44
Marketed output	–	61	–	48	41
Sovkhozy *and other public-sector farms*					
Sown area	2	9	11	49	54
Marketed output	–	12	–	40	49
Private farms and personal allotments					
Sown area	97	13	6	3	3
Marketed output	–	27	–	12	10

Sources: TsSU SSSR, *Narodnoe khoziaistvo SSSR, 1922–1972* (Moscow, 1972), 227, 240; TsSU SSSR, *Narodnoe khoziaistvo SSSR v 1985 g.* (Moscow, 1986), 190, 207.
Notes: The *kolkhoz* (*kollektivnoe khoziaistvo*) was a co-operative farm. The land was nationalized, while reproducible assets belonged to the member households, who received the farm's net income. The farm was run by an elective management. The *sovkhoz* (*sovetskoe khoziaistvo*) was a State farm. Land and assets were nationalized. The farm was run by an appointed management and salaried worker-employees as a public-sector enterprise. Until 1929 the private sector was composed mainly of peasant farms. After 1929, it was reduced to the household allotments of collective farmers, in the first place, and also of other citizens who retained the right to a small personal allotment.

Marketed output is less than total output by the amount of on-farm consumption. A much higher, but still declining, share of the private sector in total (marketed and non-marketed) agricultural output in the 1960s and 1970s was reported by G. Shmelev, '*Obshchestvennoe proizvodstvo i lichnoe podsobnoe khoziaistvo*', *Voprosy ekonomiki*, no. 5 (1981), 69, as follows: 1960–35.6 per cent, 1965–32.5 per cent, 1970–29.7 per cent, 1975–28.3 per cent, 1979–26.5 per cent.

importance; this reflected more a closing of the gap between private and collective rewards than direct repression of private economic activity, though the latter was reported from time to time. Another aspect of restructuring was the rise of the *sovkhoz* (nationalized farm). In Stalinist ideology, the *sovkhoz* was a higher form of organization than the *kolkhoz* which was 'only' a co-operative, and there were periods both under and after Stalin when policy encouraged absorption of existing *kolkhozy* into the public sector. The major vehicle for expansion of *sovkhoz* activity, however, was the extension of the margin of cultivation into the 'virgin lands' of the interior; the new farms created there were normally *sovkhozy*.

Something which directly affected the *kolkhoz* itself was its progressive 'statisation'. As the status of the village and farm work-force improved, the *kolkhoz* became more and more like a *sovkhoz*. This was reflected in a variety of trends. *Kolkhoz* managers were no longer a mixture of ill-educated peasants and political cadres who knew nothing of farming, and collective farm management became increasingly professional and specialized. The rising status of the ordinary *kolkhoz* workers was reflected in the introduction of a minimum income based on *sovkhoz* piece-rates; with this reform, the peasant ceased to be the residuary claimant on food supplies. Other changes associated with rising status ranged from the institution of retirement pensions for *kolkhoz* workers to restoration of the automatic right to an internal passport, which brought increased freedom of off-farm movement (although without the right to a share in farm equity).

Lastly, agriculture in general and the *kolkhoz* specifically were affected by structural transformations at work in the economy as a whole. Production processes became increasingly integrated, horizontally and vertically. There was a growth of inter-*kolkhoz* and 'agri-business' organizations, reflecting economy-wide trends towards more integrated and more large-scale corporate forms. Independent *kolkhozy* came more and more frequently together to form joint enterprises specializing in livestock rearing, rural construction and secondary processing of farm products; each *kolkhoz* contributed a share of the equity and took a share of the profits.

Input Mobilization and its Results

What were the results of this restructuring of the agrarian economy? Results could be measured first of all in the achievements of Soviet food policy which now aimed at rapid improvement of the Soviet diet. And Table 9.3 shows that, after the war, the Soviet diet did improve markedly although, by Western standards, there was an unhealthy preoccupation with increased consumption of high-cholesterol, high-sugar items. But improved diet was

Table 9.3 Annual food consumption of the Soviet population: selected years, 1926/27–87 and 1990 plan (kilograms per head)

	1926/27	*1950*	*1970*	*1987*	*1990 plan*
Meat, fats	40[a]	26	48	67	70
Milk, dairy products	–	172	307	363	330–40
Eggs (units)	–	60	159	268	260–66
Fish	–	7.0	15.4	17.2	19
Sugar	–	11.6	38.8	42.5	–
Potatoes	185	241	130	98	110
Vegetables	–	51	82	95	126–35
Vegetable oils	–	2.7	6.8	10.4	10.2
Fruits	–	11	35	55[b]	66–70
Cereals	185[c]	172	149	129	135

Sources: The October 1926 and February 1927 consumption of urban manual and non-manual worker households, and of rural households of the grain surplus and deficit regions, are given in *Sel'skoe khoziaistvo, 1925–1928* (Moscow, 1929), 402–5, 408–11. For 1926/27 I show the unweighted mean of these figures, except that rural households are accorded a weight of 85 per cent and urban households a weight of 15 per cent. For later years, *see* TsSU SSSR, *Narodnoe khoziaistvo SSSR, 1922–72* (Moscow, 1972), 372; Goskomstat SSSR, *Narodnoe khoziaistvo SSSR v 1989 g.* (Moscow, 1990), 118. Figures for '1990 plan' are those of the 'Brezhnev' food programme adopted in 1982.
Notes: [a] Meat only; [b] 1988; [c] wheat and rye flour.

based on domestic supply only in part because production did not keep pace with requirements. The 1970s saw the beginning of a turn towards large-scale imports, mainly of meat and animal feedstuffs, to support rising domestic meat consumption.

The failure of agricultural production to keep pace with domestic needs was certainly not for want of resources. It is true that, as Table 9.4(a) shows, for many years the Soviet agricultural work-force had been in decline, both in absolute numbers and in proportion to the total working population. Over the same period, however, from 1940 to 1970, the number of trained agronomists, animal specialists and veterinarians multiplied from 34,000 to nearly half-a-million. Moreover, Table 9.4(b) suggests that declining labour supplies were hugely compensated for by the increase in capital investment in agriculture which rose, not only in billions of 'comparable' roubles, but even in proportion to total investment in the Soviet economy. This, more than anything, indicated the rising priority of agriculture for Soviet decision-makers.

Table 9.4 Inputs into Soviet agriculture: selected years, 1940–89
(a) Employment in agriculture

	Millions	*Per cent of total work-force*
1940	28.1	54
1950	27.9	48
1970	24.1	32
1989	19.7	19

Source: TsSU SSSR, *Narodnoe khoziaistvo SSSR, 1922–1972* (Moscow, 1972), 283, 343; Goskomstat SSSR, *Narodnoe khoziaistvo SSSR v 1989 g.* (Moscow, 1990), 46, 520.

(b) Investment in agriculture

	Billion roubles, p.a., at 'comparable' prices	*Per cent of total investment*
1940	0.8	11
1956–60	5.3	14
1966–70	13.3	17
1981–5	31.2	19
1986–9	28.6	17

Source: Goskomstat SSSR, *Narodnoe khoziaistvo SSSR v 1987 g.* (Moscow, 1988), 294; Goskomstat SSSR, *Narodnoe khoziaistvo SSSR v 1989 g.* (Moscow, 1990), 533.

An independent Western assessment of the efficiency with which the growing volume of inputs was utilized is reported in Table 9.5. This table gives rise to a mixed evaluation. In the 1950s and 1960s, the growth of agricultural output in the USSR substantially exceeded the United States' record. The gap was essentially due to the higher rate of increase of Soviet inputs, including a higher rate of retention of farm-workers in agriculture, for, despite heavy Soviet investments, the rate of increase in capital intensity of production was actually more rapid in the United States. Nonetheless, in the outcome, there was little to choose between Soviet and American agriculture in terms of dynamic efficiency; in both countries, multifactor productivity in agriculture rose yearly by about 2 per cent in the 1950s and 1 per cent in the 1960s.

But what this meant was that, because the level of output per worker and of multifactor productivity was far higher in the United States to begin with, the static efficiency gap did not close. What was worse, in the 1970s, the dynamic efficiency of American farming improved a little (though not recovering the rate of improvement of the 1950s), while that of Soviet

Table 9.5 Economic growth of agriculture: USA and USSR, 1951–77 (per cent p.a.)

		1951–60	*1961–70*	*1971–77*
USA	output	2.1	1.1	2.6
	inputs	0.1	0.0	0.9
	of which, labour	−4.2	−5.0	−3.3
	input productivity	2.0	1.1	1.7
	of which, labour	6.7	5.9	6.1
USSR	output	4.8	3.0	2.0
	inputs	2.7	2.1	1.6
	of which, labour	−0.6	−0.4	−1.5
	input productivity	2.1	1.0	0.4
	of which, labour	5.4	3.4	3.5

Source: Douglas B. Diamond and W. Lee Davis, 'Comparative growth in output and productivity in U.S. and U.S.S.R. agriculture', in *Soviet Economy in a Time of Change*, vol. 2 (1979), 32, 38.

farming deteriorated further. In this, trends in Soviet agricultural production tended to mirror processes at work in the economy as a whole, summed up in a huge Soviet and Western specialist literature as relative economic retardation and stagnation.[16]

Thus, as Soviet power entered its final decade, it remained the case that an agricultural work-force, proportionally much larger than that deployed in the United States, continued to feed the domestic population at a dietary level which, despite absolute improvement, remained relatively lower. Moreover, Soviet agriculture achieved this only with the help of United States' food surpluses left over after American consumers had eaten their fill.

Problems of Resource Management

Behind the disappointing performance of Soviet supply lay profound problems of resource management. But it is worth saying at the outset that, by the 1970s, these no longer included in any prominent way the special legacy of the countryside from the Stalin years, the problems of low morale and lack of incentive resulting from the brutal suppression of peasant interests in the 1930s. The very low and uncertain return to work in the collective sector, the systematic coercion of *kolkhoz* labour – these were no longer central to the agricultural problem. In the 1950s and 1960s, such problems were substantially mitigated, mainly by raising government procurement prices (although, because official consumer prices were held down, this course carried a high price tag in terms of the rising budgetary subsidy of farm incomes).

Low morale and lack of incentives certainly persisted in agriculture but, increasingly, these reflected simply the problems of the Soviet economic system as a whole, and no longer any special historical circumstances of the rural economy.[17] Such problems can be summarized conveniently in the list that follows.

Farm-workers' lack of interest in results

This was promoted by the payment system. The payment system for collective farm-workers, like that for State farm-workers and public-sector employees generally, still meant reward according to labour input, not output. This applied whether we think of the traditional form of payment on the *kolkhoz* – the work-point (trudoden') system, which allocated to the farm-workers a share of the farm's residual net income in proportion to work done, or the more recent form of minimum payments based on task rates, which potentially broke the link between output and reward altogether.

An irrational structure of procurement prices

After abandonment of the Stalinist policy of near confiscation of food surpluses, government procurement prices were progressively raised, on average, to cover farm production costs. Within the aggregate, however, crop prices were raised substantially above production costs so that the arable sector became profitable, while livestock farming continued to incur losses. Left to themselves, net-income-maximizing collective farmers would have tended to abandon animal husbandry and leave meat production to the private sector, concentrating on crop raising. Therefore, because the State continued to require a *kolkhoz* livestock sector, farm-managers were not, in fact, left alone to maximize a surplus but, instead, remained subject to constant direction and correction of decisions from above, from the various ministerial bureaucracies concerned with agriculture and procurements.

Pressure from above for quick results

The policies promoted by ministerial officials to solve problems of food supply were systematically biased towards short-term results; this pressure on farm management resulted in short-term and in long-term misallocations. Several examples may be given. Periodic campaigns for an immediate increase in meat sales tended to result in premature slaughter and stock losses. Under conditions of uncertain rainfall, pressure for dramatic harvest improvements promoted harvest instability; the soil suffered from an inadequate moisture reserve, because too much plough-land was regularly sown and harvested, leaving an insufficient allocation of crop area to fallow. The option of extending the margin of cultivation into virgin lands, when available, nonetheless functioned to postpone consideration of necessity measures to raise efficiency in using existing inputs, rather than increase inputs further.

Over-centralization of supply of inputs, and targets for output

Output targets for individual farms were fixed from above by officials remote from village reality, ignorant of local resources, conditions and possibilities for rational specialization. Input allocations were similarly determined, resulting in uncertainty of often inappropriate supplies. The machine technology provided would frequently turn out to be inappropriate; the infrastructure of transport services, machinery parts, food storage and agronomic support was certainly inadequate.

Summary

In the past, Soviet agriculture suffered from specific problems. These problems were rooted in the Tsarist and Stalinist models of mobilizing resources out of agriculture. In the Khrushchev and Brezhnev years, a new model, based on mobilizing resources into agriculture, solved some of the old problems. The Stalinist legacy was overcome, and agricultural problems began more and more to resemble those of the economy as a whole. The defects of the economy as a whole, however, were nowhere more embarrassingly visible than in agriculture.

In the mid-1980s, under a new General Secretary, the idea of a fundamental departure from the existing model returned to the fore. Under the stimulus of the Chinese decollectivization and transition to State tenancy after 1979, the debate over the possible role of the *zveno* was renewed. Gorbachev himself rejected the idea of a hierarchy of ownership from individual peasant agriculture through the *kolkhoz* to the *sovkhoz*. Moves were initiated towards long-term subcontracting of basic farming tasks from the *kolkhoz* to the small co-operative unit, with value of output as the basis of reward, the *kolkhoz* becoming no more than a means of large-scale co-operative supply of inputs and marketing of outputs.[18] In Gorbachev's conception, which deliberately evoked the form of urban-rural exchange previously established in the 1920s, the State would contract with the *kolkhoz* for a fraction of farm output, the rest being delivered through voluntary marketing.[19]

Although these measures were very radical by Soviet standards, they were never likely to revolutionize the performance of the Soviet agricultural system; for two reasons. First, no restructuring of rural institutions was going to give good results while the economic system as a whole remained insensitive to the needs of the village community. Second, as a growing current of Soviet radical reform opinion recognized, renewal of the agrarian economy, like that of the economy as a whole, required more than economic change. It also required a redistribution of responsibilities and rights. Rural producers would have to make the difficult transition from passively surrendering food surpluses and receiving supplies to the responsible exercise of power, with equal citizens' rights of participation in, and control over, the fate of the rural society and ecology.[20] Whether the collapse of Soviet power at the end of 1991 would eventually make such a redistribution of power possible remains to be seen.

Notes

1 Alexander Gerschenkron, *Economic Backwardness in Historical Perspective* (Cambridge, Mass., 1962), Ch. 6 ('Russia: patterns and problems of economic development, 1861–1958'); W. W. Rostow, *The stages of economic growth* (New York, 1970), 66.
2 James Y. Simms, 'The crisis in Russian agriculture at the end of the nineteenth century: a different view', *Slavic Review*, vol. 36, no. 3 (1977).
3 Paul R. Gregory, 'Grain marketing and peasant consumption, Russia, 1885–1913', *Explorations in Economic History*, vol. 17, no. 2 (1980).
4 John Thomas Sanders, '"Once more into the breach, dear friends": a closer look at indirect tax receipts and the condition of the Russian peasantry, 1881–1899', *Slavic Review*, vol. 43, no. 4 (1984).
5 Paul R. Gregory, *Russian National Income, 1885–1913* (Cambridge, 1982), 170–5.
6 R. W. Davies, *The Development of the Soviet Budgetary System* (Cambridge, 1958), 65. Current roubles of 1924/25 are deflated for comparison with 1913 according to a wholesale price index (ibid., 89).
7 On the village retention of food surpluses, and the reasons for it, *see* Mark Harrison, 'The peasantry and industrialisation', in R. W. Davies (ed.), *From Tsarism to the New Economic Policy: continuity and change* (London, 1990), 109–17.
8 On the collectivization process, *see* R. W. Davies, *The Industrialization of Soviet Russia*, vol. 1, *The Socialist Offensive: the Collectivization of Soviet Agriculture, 1929–1930* (London, 1980).
9 S. G. Wheatcroft and R. W. Davies, 'Population', in R. W. Davies, M. Harrison, and S. G. Wheatcroft (eds), *The Economic Transformation of the Soviet Union* (Cambridge, 1994), 74–6.

10 V. P. Danilov, *'Diskussiia v zapadnoi presse o golode 1932–1933 gg. i. "demograficheskoi katastrofe" 30–40–x godov v SSSR'*, *Voprosy istorii*, no. 3 (1988), 117.

11 For a recent survey of general factors affecting the trend of agricultural production in the inter-war years, *see* S. G. Wheatcroft and R. W. Davies, 'Agriculture', in Davies, Harrison, and Wheatcroft (eds), *Economic Transformation*, 106–130.

12 Davies, Harrison, and Wheatcroft (eds), *Economic Transformation*, 290.

13 On the origins of the *kolkhoz*, *see* R. W. Davies, *The Industrialization of Soviet Russia*, vol. 2, *The Soviet Collective Farm, 1929–1930* (London, 1980).

14 On the *zveno* debate in 1939 and 1950, *see* Leonard Schapiro, *The Communist Party of the Soviet Union*, 2nd edn (London, 1970), 520.

15 For successive overviews of Soviet agricultural policy and progress under Khrushchev and Brezhnev *see* articles in the triennial collections published by the United States Congress Joint Economic Committee, especially David W. Carey, 'Soviet agriculture: recent performance and future plans', in *Soviet Economy in a New Perspective* (Washington, D. C., 1976); David W. Carey and Joseph F. Havelka, 'Soviet agriculture: progress and problems', and David M. Schoonover, 'Soviet agricultural policies', both in *Soviet Economy in a Time of Change*, vol. 2 (Washington, D. C., 1979); D. Gale Johnson, 'prospects for Soviet agriculture in the 1980s' in *Soviet Economy in the 1980s: Problems and Prospects*, vol. 2 (Washington, D. C., 1982).

16 For an accessible summary of this literature, *see* Paul R. Gregory and Robert C. Stuart, *Soviet Economic Structure and Performance*, 4th edn (New York, 1990), Ch. 12.

17 Addressing a conference in Washington, D. C., in 1981, the American scholar, Gertrude Schroeder, remarked: 'What's wrong with Soviet agriculture? What's wrong with Soviet agriculture is that it's part of the Soviet economy.'

18 Karl-Eugen Wädekin, 'The re-emergence of the kolkhoz principle', *Soviet Studies*, vol. 41, no. 1 (1989), 35.

19 This was noted by R. W. Davies, *Soviet History in the Gorbachev Revolution* (London, 1989), 28.

20 Teodor Shanin, 'Soviet agriculture and perestroika: four models', *Sociologia Ruralis*, vol. 29, no. 1 (1989), 15.

Notes on Contributors

Paul Corner is Professor of History at the University of Siena, Italy. He is author of *Fascism in Ferrara* (Oxford University Press, 1974), and his recent publications include *Contadini e industrializzazione* (Laterza, 1993), and, with A. Cento Bull, *From Peasant to Entrepreneur: The Survival of the Family Economy in Italy* (Berg Press, 1993).

John A. Davis was formerly Director of the Centre for Social History at the University of Warwick and is currently holder of the Emiliana Pasca Noether Chair in Modern Italian History at the University of Connecticut. He is joint editor of the *Journal of Modern Italian Studies* and is preparing the volume on Italy for the *Oxford History of Modern Europe*.

Mark Harrison is Reader in Economics, University of Warwick. Author of *Soviet Planning in Peace and War. 1938–1945* (1995), *The Soviet Home Front. 1941–1945* (with John Barber), (1991), *The Economic Transformation of the USSR 1914–1945* (joint editor with R. W. Davies and S. G. Wheatcroft), (1994), articles in *Journal of Economic History, Economic History Review, Soviet Studies/Europe-Asia Studies*, and other journals.

Colin Heywood is Senior Lecturer in Economic and Social History at the University of Nottingham. He is the author of *The Development of the French Economy. 1750–1914* (London, 1992).

B. A. Holderness is Senior Lecturer in Economic History at the University of East Anglia. His publications include *Pre-industrial England: Economy and Society. 1500–1750* (Rowan and Littlefield, 1976), *British Agriculture Since 1945*

(Manchester, 1985), and, with M. Turner, has edited *Land, Labour and Agriculture 1700–1920* (1991).

Peter Mathias is former Chichele Professor of Economic History at Oxford 1968–87 and the Master of Downing College, Cambridge from 1987–95. His main publications include *A History of the Brewing Industry in England, 1700–1830* (1959 and 1993), *The First Industrial Nation* (1969 and 1983), and *The Transformation of England* (1979). He is a Fellow of the British Academy, Hon. President of the International Economic History Association and Vice-president of the 'Datini' International Institute at Prato, Italy.

Mark Overton is Professor of Economic and Social History at the University of Exeter. He has published extensively on the agrarian history of England and his *Agricultural Revolution in England: the Transformation of the Rural Economy 1500–1850* is published by Cambridge University Press in 1996.

Roger Price is Professor of Modern History at the University of Wales, Aberystwyth. His recent publications include *A Social History of Nineteenth Century France* (1987), *The Revolution of 1848* (1988) and *A Concise History of France* (1993). He is currently writing a study of *The French Second Empire: State and Society* and completing a programme of research into technical innovation in the late eighteenth and nineteenth centuries which focuses on industrial power sources.

Kaoru Sugihara is Senior Lecturer in Economic History of Japan at the School of Oriental and African Studies, University of London. He has written extensively on modern Asian international economic history, and co-edited *Local Suppliers of Credit in the Third World, 1750–1960* (Macmillan, 1993) and *Japan in the Contemporary Middle East* (Routledge, 1993).

F. M. L. Thompson is former Director of the Institute for Historical Research in London University. He is author of *English Landed Society in the Nineteenth Century* (1963), *Cambridge Social History of Britain 1750–1950* (1990), *The Rise of Respectable Society: A Social History* (1988), and his recent publications include an edited volume, *Landowners, Capitalists and Entrepreneurs* (1994).

David Washbrook is Reader in Indian History at the University of Oxford and a fellow of St Antony's College. He is author of *The Emergence of Provincial Politics: the Madras Presidency. 1870–1920* (Cambridge, 1976), and has published widely on Indian nationalism and economic history.

Index